BOOKS OF SECRETS

BOOKS OF SECRETS

Natural Philosophy in England, 1550–1600

Allison Kavey

UNIVERSITY OF ILLINOIS PRESS

URBANA AND CHICAGO

© 2007 by Allison Kavey
All rights reserved
Manufactured in the United States of America
C 5 4 3 2 1
♾ This book is printed on acid-free paper.

Library of Congress Cataloging-in-Publication Data
Kavey, Allison, 1977–
Books of secrets : natural philosophy in England,
1550–1600 / Allison Kavey.
p.   cm.
Includes bibliographical references and index.
ISBN 978-0-252-03209-7 (cloth : alk. paper)
1. Philosophy, English.
2. Philosophy, British. I. Title.
B1111.K38   2007
082.0942—dc22   2007011400

To Andrea.
This book is dedicated to the unfolding of our secrets.

CONTENTS

ACKNOWLEDGMENTS ix

INTRODUCTION
Telling Secrets 1

CHAPTER ONE
Printing Secrets 9

CHAPTER TWO
Roger Bacon, Robert Greene's Friar Bacon, and the Secrets of Art and Nature 32

CHAPTER THREE
Structuring Secrets for Sale 59

CHAPTER FOUR
Secrets Gendered: Femininity and Feminine Knowledge in Books of Secrets 95

CHAPTER FIVE
Secrets Bridled, Gentlemen Trained 126

CONCLUSION
A Secret by Any Other Name 156

NOTES 161

BIBLIOGRAPHY 187

INDEX 195

ACKNOWLEDGMENTS

The first and biggest thank you must go to Larry Principe. Without his sense of humor and unflinching support, even in the face of literary theory, this project would have looked very different. He should be hailed as a scholar and a gentleman.

Thanks to the institutions that have supported this work, in particular the Department of the History of Science, Medicine, and Technology at the Johns Hopkins University. Thanks to the sadly disbanded Center for Research on Culture and Literature at Johns Hopkins, which funded a year of research that came complete with an excellent seminar and demonstrated the strengths of interdisciplinary scholarship. Thanks to Clare Carroll and the other members of her Folger Library seminar, "Genealogies of Britishness," who have been a source of camaraderie and inspiration since our first meeting.

Thanks to the staff at the Folger Library, the British Library, the Bodleian Library, the Huntington Library, Guildhall Library, the Wellcome Institute for the History of Medicine, Special Collections at the Eisenhower Library at Johns Hopkins, and Special Collections at Northwestern University for their help in finding my "dwarf books."

Thanks to Joan Catapano, Kerry Callahan, and Kristen Ehrenberger and the rest of the editorial staff at the University of Illinois Press who have supported this book since its very early days.

Thanks to my family, whose consistent optimism and support have made this, and everything else, possible. Thanks to my friends, who have weathered a great deal of drama and nonsensical babble with good spirits. Thanks especially to Mark, who was there every day and listened to me crow and cry for years with exquisite good humor. Thanks to Jerry, who reminds me every day about the unpredictability of nature, and who could not care less about this book unless it guarantees his supply of marshmallow chickens. Thanks to Nikki, around whom nature revolves, and to Thursday, the personality behind the preternatural.

BOOKS OF SECRETS

INTRODUCTION

Telling Secrets

"Humans have the most amazing talent, a special kind of stupidity.
You think the whole universe is inside your heads."
—TERRY PRATCHETT[1]

This project began as a quest for the meaning and cultural status of books of knowledge, inexpensive perpetual almanacs that included information on astrology, physiognomy, and medical theory and advice. The original goal was to consider the ways in which these books helped readers reimagine the world around them as predictable and explicable, rather than opaque and chaotic. In the process of researching the works of Erra Pater, the famed wandering Jew of early modern Europe, I stumbled upon two things that sent me in another direction. The first centered on Erra Pater himself—it seemed important to understand what this clearly mythical character from the classical knowledge tradition meant to early modern readers and why he was chosen by printers to be the "author" for the series of books that bears his name. The second was a line from the introduction to Erra Pater's *Book of Knowledge,* which described its contents as suitable for "readers of the meanest capacity." Suddenly the question of who was expected to, as well as who actually did, read and use these books as scripts to interpret their worlds became central to interpreting their utility and position in the cheap print market.

Books of secrets offer an excellent avenue for gaining insight into both these questions, as well as providing an opportunity for considering themes shared by popular constructions of the natural world. These small books, like books of knowledge, occupy a significant place in the marketplace of inexpensive print that burgeoned in the second half of the sixteenth century in England. They also comprise a clearly defined genre, are attributed to a wide variety of ancient and contemporary authors, and both directed their content to and held intellectual appeal for a broad audience that included aristocrats, academics, and "readers of the meanest capacities." Finally, and most appealing to me, they offered the opportunity to think about the meaning and cultural currency of secrets, a tantalizing and inevitably protected kind of knowledge. This work will focus on the position these books held in that marketplace. It will interrogate the meaning of secrets in these books, ask who wrote, printed, and read them, and examine the kinds of worlds they presented to readers.

William Eamon remains an important authority on secrets and science in early modern Europe. His work links European sixteenth-century books of secrets with the development of the "Baconian sciences" like chemistry, metallurgy, magnetism, and electricity and their departure from the theoretical traditions that informed scholastic discussions of the natural world.[2] He agrees with Thomas Kuhn's theory of the scientific revolution that the experimental practices associated with these avenues of investigation contributed to the formalization of scientific method and investigation.[3] He also argues that the popularization of the tradition of secrets, a process he traces from its medieval roots through its elite reception in the Renaissance and eventual peak in the sixteenth century, laid the groundwork for the exposure of nature's secrets that composed seventeenth- and eighteenth-century scientific practice.[4]

Eamon's work is significant because it provides a survey of the genre of books of secrets and their position in sixteenth-century European culture. His book is noteworthy for its extensive investigations of notable books of secrets and the people behind them, particularly Ruscelli's edition of *The Secretes of Alexis of Piemonte*.[5] It is also important for its consistent efforts to pin down the social significance and meaning of secrets in Europe, particularly Italy. Eamon's book remains problematic, however, since its central arguments compel him to accept the scientific revolution as given and search for "the prehistory of the Baconian sciences" and the "missing link between medieval secrets and Baconian experiments."[6] Eamon's definition of popularization and the audience he believes to have been targeted by these books is also difficult to determine, since he does not consider the range of prices for the books he discusses nor the audiences to which they were advertised.

This book addresses the questions left unanswered by Eamon's *Science and the Secrets of Nature*. It considers books of secrets that are included in the group of books classified as "cheap print," meaning that they cost six pence or less, and it locates them within popular print culture, an area that has received significant attention since Eamon's book was released. It begins with an intensive analysis of the printers who produced books of secrets and concludes that those printers were linked through bonds of apprenticeship and patronage that differentiated them from other men working in the print trade. It focuses on books that produce coherent models of the natural world and differentiates them from the discipline-specific books of secrets that came out of various craft traditions.[7] It builds on Adam Fox's recent work on the connections between oral and print culture to argue that these books could operate on both levels, a point demonstrated best by Gervase Markham's book of secrets, the focus of chapter 5 of the present work.[8]

The remainder of the book concentrates on the knowledge that these books made available to a wide range of readers. It argues that this knowledge, classified as natural because it involved natural phenomena, the forces governing natural change, and natural objects, made the inner workings of the natural world accessible to people who lacked access to academic natural philosophy. It contends that these books presented readers with models of the natural world that were susceptible to human manipulation, pointed to the agents of natural change that were vulnerable to such efforts, and issued instructions for producing particular changes that would suit or satisfy human desires. Finally, it argues that the effect of all these books was to authorize readers' perceptions of the natural world and recognize their position, and sometimes even centralize their position, in it.

One of the central themes explored in the following chapters is the construction of authority in books of secrets. Authority exists on at least two levels, often three, in these small books, and the work dedicated to establishing the meaning of authority on all these levels is worth investigation. The first person to acquire authority in each of these texts is the author, a term that has to be defined generously to have meaning for this group of books. In the case of *The Secrets of Albertus Magnus* and *The Mirror of Alchimy*, the "authors," Albertus Magnus and Roger Bacon, were medieval religious men famous for their mastery of the natural world. The books attributed to them reflected different aspects of their natural philosophical work. In the case of Albertus Magnus, the knowledge in the book mirrors his nickname, Doctor Universalis, since it includes the virtues of stones, beasts, and plants and a catalog of wonders. It also identifies the systems governing the natural world, specifically sympathy and antipathy. The two treatises attributed to Bacon in *The Mirror of Alchimy*, on the other hand, reflect his specific reputation for stretching the limits of art and nature. It also presents his version of alchemical theory, which includes recommendations for manipulating the world by mimicking natural systems. Ensuring the authority of these two men was more an act of reminding readers of their names and titles, particularly their academic and religious accomplishments, than of creating any authority for them.

Established authority coexists with and is supported by readers' experience in *Cornucopiae, Or divers secrets*.[9] That book offers an image of the natural world in which readers' observations contextualize and make accessible broader theoretical and academic constructions of nature. Readers are also encouraged to accept the wisdom and observations of the compiler, who guarantees the validity of some of the information he presents by claiming to have witnessed it himself. In Gervase Markham's work, a living author's experience is introduced as superior to the accepted wisdom of ancient authorities. In order to establish his authority, he gives specific examples in which he followed accepted wisdom and lost a great deal of time and money. Then he offers a new approach to these problems, assuring his readers that his solution works, although it contradicts inherited wisdom.

Steven Shapin has laid the groundwork for considering the construction of scientific authority in seventeenth-century England. He argues that anyone identified as a gentleman, a title determined by blood lineage and income, was considered a good witness, someone whose experience was reliable. He bases this claim on the early modern belief that gentlemen were not beholden to anyone, and so their judgment would not be clouded by personal relationships, and that gentlemen's skills of observation were heightened because they did not waste energy and dull their nerves on labor required to support themselves.[10]

Shapin's version of authority does not extend to women, even gentlewomen. He notes that women's inherent dependence, which is derived from their status as daughters and wives, disqualifies them from being good witnesses, since their observation could be skewed by their personal debts.[11] I argue that *The Widowes treasure*, examined here in chapter 4, challenges this notion, since the knowledge it presents was originally the product of a woman's medical practice and certainly represents her household experience. A similar example of female authority can be found in the gentlewoman who urges the printing of the two books compiled by James Partridge, *The Treasurie of commodious Conceits, and hidden Secrets* and *The Treasurie of hidden Secrets*. These constructions suggest that early modern women, at least women of a certain social standing, did have the authority to guarantee the utility of a book of secrets. Bernard Capp's examination of female authority in the household and in relationships to men may offer the best model for developing a broader definition of authority that includes women.[12]

Steven Orgel offers another conception of authorship developed from several decades of scholarship in English literature and drama that emphasizes the importance of multiple authors in early modern print, particularly plays.[13] He argues that authority coalesces around the name on the title page, but that the actual writing of plays and books during the early modern period was a group effort. I would like to extend his argument to include books of secrets, particularly those compilations that reflect multiple intellectual traditions and the efforts of at least one compiler and printer. One of the best examples of this can be found in the attribution of books of secrets to ancient or medieval authorities who had absolutely nothing to do with them, but whose reputations helped to guarantee the value of the knowledge in the books and acted as a selling point. This will be explored at length in chapter 2.

Authors, compilers, and printers were not the only people to have authority in these books. Readers were also granted status as witnesses and judges of useful natural knowledge. The empowerment of readers, particularly since the socioeconomic range of people who could have purchased and read these books was broad, is the best evidence that Shapin's notion of authority is too narrow, both temporally and socially, and needs revision. While he acknowledges the potential for common people to occasionally transcend their class limitations, he is reluctant to offer authority to anyone whose sex or social class compromises their independence or their ability to serve as a reliable witness.

All these books take the same approach to developing readers into authorities. They begin by defining the characteristics that make a good reader of secrets, namely diligence, perseverance, and carefulness, and then assert that only those readers would be capable of making use of the natural knowledge contained in the books. They continue with a series of strategies, including emphasizing the importance of readers' experiences of and in the natural world, the place they and their needs occupy in relationship to the world, and their perceptions of their ability to manipulate or control nature. Readers have, by the conclusion of these books, gained a sense of themselves as practitioners, as authorities, and as active participants in, rather than victims of, the world around them. Chapters 2 through 5 explore the ways in which books of secrets created a definition of desirable readers and authorized readers' observations of and experiences with the natural world.

In order to generate books of secrets that recognize readers' needs and experiences, the natural world has to be depicted as vulnerable to human manipulation. All the books of secrets examined in this work depict worlds in which the category of the natural has been expanded to include hidden forces, which had previously been considered preternatural and outside the grasp of legitimate art. The dissolution of this category and a nuanced conception of the way people understood the meaning of "natural" has been best covered by Lorraine Daston, who argued that the category of preternatural was losing utility by the seventeenth century and was being replaced by an expanded conception of "natural" action and phenomena.[14]

I believe that this trend is echoed in books of secrets, where it has two effects. First, readers are presented with worlds governed by predictable forces, particularly sympathy and antipathy, which govern relationships and dictate change. Second, practitioners of natural philosophy are given the opportunity to extend their legitimate inquiries into the workings of occult forces in the natural world. Since the category of practitioners has been enlarged in these books to include people from a broad range of social, economic, and academic backgrounds, the expansion of the natural extends the power to manipulate the world to a large number of people.

The people responsible for making secrets available to a broad population for the first time are the printers who proliferated in the second half of the sixteenth century in London. Chapter 1 reviews the creation of the Stationers' Company and then offers an analysis of the different ways in which printers made use of and understood books of secrets. It argues that each of the books discussed here represents a different comprehension of the importance of cheap print in the print marketplace, the positioning of natural knowledge in popular culture, the intricacies of printers' networks, and the connections between printers and authors.

Chapter 2 examines the importance of adopting a medieval natural philosopher, Roger Bacon, known for his extensive knowledge of the natural world of secrets. This chapter charts the interactions between the theater and the print shop, arguing that Robert Greene's play *The honorable historie of frier Bacon and frier Bongay* capitalized on Bacon's aristocratic and academic currency in sixteenth-century England and presented him and his dramatic power over the natural world to a broad audience of London theatergoers. The success of the play, which saw many productions between 1591 and 1593 in London and most likely played the countryside as well, encouraged the printing of the quarto edition of the script, which shifted Bacon from the boards to the page, where he stayed alive as a character in popular culture. A book of secrets attributed to Bacon appeared in 1597, and it emphasized a particular aspect of his popular characterization, both on the stage and on the page. It depicted Bacon as an extensively educated man whose study of the natural world had led him to stretch, and then rewrite, the boundaries of art and nature. His presentation as a dramatic character who tests the outer limits of art and nature is echoed in the final treatise of *The Mirror of Alchimy*, which attributes to him a lengthy discussion reassigning manipulation of occult forces to natural philosophy and dismissing magic as the tricks of charlatans intended to fool the ignorant.

One of the best ways to access the ways in which these books work is to examine their forms. This approach owes much to literary criticism, which pioneered formalism as a useful strategy for understanding texts. Literary historian Richard Helgerson defends his own use of this technique on sixteenth-century documents, writing "I assume—that discursive forms matter, that they have a meaning and effect that can sometimes complement but that can also contradict the manifest content of any particular work. Forms in this view are as much agents as they are structures. They make things happen."[15] I take this to heart in chapter 3, in which I investigate the importance of textual structure in *The Secrets of Albertus Magnus* and *Cornucopiae, Or divers secrets*, two books of general natural knowledge. The structure of each book works to promote a particular image of the world and its governing forces, and it also supports a specific model of readers' relationship to that world and ability to predict and manipulate it.

Chapter 4 takes up the task of investigating the mutable qualities of nature and presenting them to a broad audience in the name of telling the secrets of nature to women. The role of women in early modern England has attracted significant attention in recent years. Sara Mendelson, Patricia Crawford, and Laura Gowing, in particular, have investigated the responsibilities women had in the sixteenth and seventeenth centuries, charted their life courses, and concentrated on their relationships with popular culture, particularly popular print.[16] Jennifer Stine, Sara Pennell, and Margaret Pelling have worked on women's interactions with print, particularly popular cookery and medical manuals.[17] Chapter 4 investigates the only books of secrets in sixteenth-century England that were attributed, at least in part, to female authors or sponsors and aimed at a female audience.

Chapter 4 begins by investigating the ways in which an anonymous female practitioner and a similarly nameless gentlewoman were presented, like the long-dead Albertus Magnus and Roger Bacon, as experts in natural knowledge. It continues by examining the meaning of secrets in these books, which define legitimate natural manipulation as useful natural manipulation. It also continues to examine the importance and meaning of structure in these books, focusing on the recipe as a means for conveying natural knowledge and an image of the natural world as predictable and vulnerable to manipulation through adherence to carefully charted steps. Finally, it examines the multiple arenas in which natural knowledge has currency for female readers and practitioners, specifically the kitchen, the garden, and the sickroom, and the ways in which that knowledge contributed to a definition of early modern femininity.

Chapter 5 examines a book aimed at an audience composed of rather strange bedfellows, gentlemen and the servants who worked on their farms and cared for their horses. Gervase Markham's book of secrets is addressed to gentlemen interested in breeding, training, riding, racing, and caring for horses, but it would actually have had practical impact on the lives of the boys and young men who did the work required to care for those horses. While it presents a great deal of theoretical and practical information that would have guided gentlemen's choices in stable and pasture architecture, breeding choices, and training approach, it also contains a tremendous amount of guidance for the daily labor of caring for expensive equines. The chapter argues that stable lads and stable managers, while not the stated audience for the book, represented a good portion of the people who were supposed to read and make use of it. This interpretation has significance for mapping the transmission of printed secrets across class boundaries and points to the multiple levels of meaning contained in books of secrets. Furthermore, it contributes to two definitions of masculinity: one for gentlemen, which included having the funds and time to purchase and maintain expensive horses, and one for stable lads, whose reputations depended upon their devotion of proper care and attention to turning out exquisite equines.

All these chapters indicate that books of secrets did, in fact, make the manipulation of nature more accessible to a broad variety of people. They accomplished this by consistently depicting the world as vulnerable to human interference, giving detailed instructions for exactly how to accomplish desired changes, and including that knowledge in an affordable package. They also defined their readers as powerful agents in, rather than victims of, the natural world. Quite a bargain for six pence!

CHAPTER ONE

Printing Secrets

The second half of the sixteenth century was marked by the formalization of English printing with the 1555 incorporation of the Stationers' Company. The company proved to be a functional institution around which printers could coalesce and on whose regulations they would base an increasingly vibrant and lucrative trade. Two generations of men printed the majority of texts between the stationers' incorporation and Elizabeth I's death in 1603. The first generation included the signers of the company's incorporation papers, primarily men who already made at least some of their living from printing. They produced, reinforced, and refined the regulations that had defined printing practice in Edward's, Mary's, and the first decades of Elizabeth's reign. They also trained the men who would follow them in the print shops. The second generation of English printers began their careers as apprentices to the original members of the company, from whom they learned their trade and inherited a complicated web of social obligations and interactions that influenced the locations they chose for their shops, the books they licensed and printed, the partnerships they formed with other printers, and the apprentices they chose for themselves.

This chapter will examine the relationships that shaped the English print market by analyzing the social networks that influenced the production of books of secrets. An examination of the printers of *The Secrets of Albertus Magnus* will demonstrate the connections between printers and their books, while an analysis of the intersections surrounding *The Treasurie of commodious Conceits, and hidden Secrets* and its components and *The Widowes treasure* will reflect the connections among printers. A final set of interrogations will center on the relationship between a single author and a printer and the effects of their collaboration on both their careers and the broader popular print market. More broadly, the chapter will address the location of "secrets" in the print marketplace in an attempt to describe the place they held and the ways in which these books reflect a set of connections among printers and an agreement about the meaning and cultural position of books of secrets.

Printers and Their Books

The four sixteenth-century editions of *The Secrets of Albertus Magnus* provide an excellent place to begin this investigation, since it was the first book of secrets to be translated into English. It was printed first in 1560 with an edition by John King, then in 1565 by William Copeland, then in a 1570 edition by William Seres, and finally by William Jaggard in 1599. The book must have sold well, implying some level of resonance with readers, or it would not have been reprinted so frequently. Because of its popularity, it occupied an important position in the print marketplace, bridging the first two generations of company printers and serving as an integral part of the collections of some of the most successful London printers in the middle and late sixteenth century. An examination of the career trajectories of these men that includes an accounting of their colleagues, their positions within the Stationers' Company, and the other books they licensed will produce a richer understanding of the role that a single book of secrets played in the early decades of company printing and the different ways in which books like this one could be printed, combined with other books, and reprinted to ensure the survival of their producers.

John King, the printer of the 1560 edition under the title *The boke of secretes of Albertus Magnus,* wrote the forty-third signature on the original bill of incorporation for the Stationers' Company.[1] He printed actively between 1555 and 1561, during which time licensing fees were recorded on his behalf for eighteen books. He was fined for printing two books without a license and it is possible that he escaped the notice of his peers for producing a few more unlicensed titles.[2] King was far from the most successful printer of the first generation of company men, as demonstrated by the relatively small number of books he licensed. It is possible to more perfectly locate his financial situation by comparing his contributions to company projects with those made by his peers. In 1556, for example, he contributed four pence to a drive initiated by the Lord Mayor of London and the court of Aldermen to benefit Bridewell. John Awdelay, an apprentice who had not yet completed his term of service, matched

King's donation of four pence, and they tied for the lowest fee paid, indicating that it may have been the minimum required of company members. The highest fee paid during this drive was three shillings, four pence, the amount offered by four masters of the company, Bonham, Waye, Wolfe, and Turke, and Thomas Duxsell who lacked an office but must have made a good income.[3]

This comparison does not reflect the sum of his finances, however, as indicated by the two shillings he donated the same year to defray the costs of the company's incorporation.[4] Hugh Cottesforth paid eight pence in that drive, the lowest amount offered, while Master Cooke captured the high end with a donation of thirty-five shillings. The Masters again contributed more on average than other members of the company, with the smallest fee being ten shillings. King's two shillings placed him in the lowest quarter of donations for that drive. His relative impoverishment, or at the very least lack of disposable income, in 1556 may be attributed to the few books he had under license, and thus the small number of titles he was able to produce and sell.

The latter years of his career saw him holding many more titles, and *The boke of secretes of Albertus Magnus* was among the last five books King printed before he died in 1561. It appeared during his most productive period, which fell in the last two years of his life, and coincided with his printing other titles dedicated to exposing aspects of managing the natural world, including the final book he licensed, *The medysine for horses*. In combination with his early production of cookery and household conduct manuals, these books indicate a commitment to providing affordable vernacular guides that could increase readers' rewards from their manipulations of the natural world.[5] Some of his high productivity during this period may be due to the gradual maturation of the two apprentices King presented to the company in 1556. It is likely that they worked with him until his death, which came before their terms of service would have been completed.[6] A second explanation for his high productivity during the latter 1550s and early 1560s can be located in the increase in licensed titles that occurred throughout the company during this period. The 1560s marked the beginning of a widespread increase in the number of printers working in London and the number of titles they licensed.

William Copeland, the printer of the 1565 edition, *The booke of secretes of Albertus Magnus*, was another charter member of the Stationers' Company, and his is the fiftieth signature among the original members.[7] Copeland's father had been a successful printer and established a productive shop at the Rose Garland in Fleet Street, which passed in 1547 to his son. The proof of this inheritance can be found in each book William Copeland printed, as he marked his transition to master of the shop by adding his name at the bottom of his father's device, a rose garland.[8] He printed successfully at that address from 1547 through 1556, a period during which he also achieved a measure of financial stability. His relative liquidity is indicated by his donation of twelve pence to the Bridewell campaign and two shillings, six pence, in support of the company's incorporation, both of which occurred in 1556.[9] That year, he was also ordered by the Privy Council to deliver all his copies of Archbishop Cranmer's *Recantation* to John Cawood, one of the masters of the company, to be burnt.[10] This command, coming in the midst of Mary's reign, is not surprising, since she and her ministers carefully policed the production of Protestant literature and were constantly prepared to overturn individual privileges granted to particular printers by Edward VI and Henry VIII.[11] It is also worth noting that neither was Copeland the only printer to be so punished, nor did the remainder of his catalog reflect extreme Protestant commitments, but instead a concerted effort to license and print books that would prove to be steady sellers.

William Copeland continued to license ballads, sermons, and other small books in the second half of the 1550s, and he attracted the attention of the company three times in 1558–1559 for printing books without a license, first Bradford's *Sermons of Repentance,* then *Nostradamus,* and *The epestelles and gospelles.*[12] By 1561 he had left the Rose Garland for a new shop, the Three Crane Wharf at the Vinetree in St. Martin's Parish, and he licensed no books that year.[13] In 1562 he moved again, to Lothbury against St. Margaret's Church, and he resumed licensing and printing inexpensive books, including a quarto edition of the *Book of Virgil,* a small book attributed to Virgilius the medieval necromancer. Between 1562 and his death in late 1568 or early 1569, he licensed twelve titles and printed at least one more, since no record of a license exists for his 1565 printing of *The booke of secretes of Albertus Magnus.* His edition of this text came, like King's, toward the end of his career. He printed it along with other books of knowledge and wonders, like *An introduction to knowledge, A breaf and pleasaunte treatese of the interpretation of Dreames,* and the second to last book he licensed, *A mooste breffe treatise of the strange Wonders seen these latter yeres in the Ayer in soundry Countryes as in Germanye & c.*[14]

Both King and Copeland developed catalogs that emphasized vernacular books devoted to interpreting and influencing the natural world. Copeland in fact devoted one-third of his printing efforts between 1562 and 1568 to books that addressed the construction of knowledge and interpretation of the natural world, which suggests that these books had proved effective in attracting customers. He did not print enough books overall, or even enough books of secrets, to ensure that this system would prove financially successful. Twelve books in six years, even with the possibility of reprints of books he had already licensed, would not have provided much of a living, and he was buried at the company's expense between June of 1568 and July of 1569, indicating that his estate could not pay for his funeral.[15]

The 1570 printer of *The Boke of secretes of Albertus Magnus*, William Seres, lived a very different life from the economically marginal ones of John King and William Copeland. Like them, he was a charter member of the Stationers' Company, but he continued to rise in power, eventually holding the position of master five times. Part of his success can be attributed to the sole right to print prayer books, primers, and psalters in Latin and English he received from Edward VI in 1553, lost during Mary's reign, and then regained in 1559.[16] The market for books for religious services burgeoned under Elizabeth I, and Seres benefited from his monopoly over their production. He did not, however, enjoy his privilege without question, especially as he neared death and attempted to negotiate the successful transition of his patents to his heirs. Toward the end of Seres's life, he assigned them, along with some of his printing equipment and copies, in exchange for annual rent to Henry Denham, who joined with seven other members of the company to allow Seres to satisfy remaining printing demands.[17]

Other printers, to whom Seres had not assigned his privilege to print private prayer books, psalters, and primers, also began printing them, claiming that the queen had no right to grant special printing rights, and that the production of all good and useful books should belong to every printer. Seres responded with a letter to the Lord Treasurer in which he argued that there was significant ancient precedent for the granting of printing privileges to loyal and skillful men by princes, and that the practice should continue because uncontrolled printing, particularly of religious texts, could be dangerous. The matter was resolved in 1583 by an agreement within the company that men with privileges would contribute an allowance to support the institution and its poorer members.[18] This agreement corresponds with Seres's demonstrated commitment to the success of the Stationers' Company, to which he donated an impressive amount of money and time over the course of his career. He donated, for example, four shillings to the company's incorporation fees in 1556 and another twenty shillings that year toward wainscoting in the council chamber and a new window in the company hall.[19] He followed these financial donations with temporal ones, beginning his first term as Warden of the company in 1561.

Seres's career cannot be summarized simply as the combination of a fortunate printing privilege and a talent for politics. Between 1555 and his death around 1580, he licensed forty titles or more, at least half of which were not protected by his monopoly of personal religious books.[20] He, like King and Copeland, produced a number of books that investigated the natural world, particularly as it applied to the human body. His catalog was heavily medical, and the books he printed offered systems for comprehending and thwarting disease at a time when the medical marketplace was relatively open and competitive. In 1564, for example, he licensed *The Dyall of Agues contanynge the names in greke laten and englesshe,* a medical text that provided an epistemology and vocabulary for the commonly diagnosed disease ague.[21] He continued his production of medical works in 1567, when he printed *Serten verces in Latin by Hippocrates* and *Master Doctour Haddons workes.*[22] The first book would have found an audience among medical students, whose training was still grounded in works attributed to Hippocrates, as well as vernacular medical practitioners interested in linking their diagnoses and prescriptions to a broadly recognized and respected medical authority.

That year Seres also expanded his investment in the natural world from the management of the human body to the characteristics of an increasingly profitable and popular natural product, wine. *The nature and properties of all Wynes that are commonly used here in Englonde &c* was Seres's last contribution to natural knowledge before his edition of *The Boke of secretes of Albertus Magnus.*[23] These titles indicate an interest in producing books of natural knowledge, particularly medical texts, which would have made *Albertus Magnus* a logical addition to his catalog, since it combined an encyclopedic account of natural products with their medical, magical, and natural philosophical properties. Like King and Copeland before him, he produced his only book of secrets as the capstone of a recognizable series of titles devoted to various aspects of natural philosophy. The text correlated nicely with some of the precedents he established with his earlier titles, particularly the celebration of ancient authority, as indicated by his edition attributed to Hippocrates and this book's attribution to Albertus Magnus, and the management of disease, as shown by his books of medical recipes and practical advice. Its production also represented the first fading of one of the brightest stars of the first generation of printers, as it would mark the rising of a brighter one in the firmament of the second generation of the Stationers' Company.

William Jaggard, the 1599 printer of *The Secrets of Albertus Magnus,* was apprenticed in 1584 to Henry Denham, the man who rented William Seres's privileges and some equipment in 1574, and he was made a free member of the company in 1591.[24] He established premises in St. Dunstan's Churchyard in 1594 and licensed his first title in 1595.[25] Unlike his predecessors, who printed their first book of secrets at the ends of their careers, Jaggard licensed *The Secrets of Albertus Magnus* as his first title and reprinted it throughout his career. Jaggard, who benefited from no royal printing privileges, seems to have made his fortune from a knack for satisfying the growing demand for affordable literary, dramatic, and poetic works by popular authors. In 1599, for example, he produced *The Passionate Pilgrim,* which he attributed to William Shakespeare, despite its notable lack of Shakespearean style. In 1608 he bought the successful printing business of James Roberts, who will be treated in chapter 5 in relation to his edition of Gervase Markham's equestrian book of secrets, and who is more broadly famous for his work on Shakespeare's quartos.[26] This clever purchase of a thriving business, particularly one that printed Shakespeare, has established Jaggard's place among the foremost printers of the books studied by academics, but it does not fully describe the range of his printing enterprise.

Jaggard's talent for recognizing and satisfying popular demand can also be witnessed in his multiple editions of popular medical and natural philosophical works. Jaggard seems to have been as invested in providing books for people interested in learning new systems and vocabularies for describing and manipulating the natural world as those devoted to the stage or the great historians and poets of England.[27]

The question his collection of titles presents so beautifully is whether the readers of all these books were in fact the same people, an argument that would erode some of the claims academics have made for differences between consumers of literary titles and consumers of vernacular print. It would thus bring into question the categories of "elite" and "common" print culture.[28] First, a definition of "popular print" seems in order. A review of the Stationers' Registers points to the kinds of books printers produced most frequently, namely ballads, pamphlets, and inexpensive collections of a wide range of types, including wonder tales, conduct manuals, and religious texts. A similar perusal of the English Short Title Catalog reflects that many of these books were produced numerous times or in multiple editions, revealing the staying power of some of this inexpensive print, including books of secrets. This kind of print constituted the majority produced by sixteenth-century printers, and the evidence from the Stationers' Registers of printers' long, often lucrative careers indicates that printers produced this kind of book because people bought it. The trend toward inexpensive editions of poetry and plays witnessed in Jaggard's catalog, particularly over the early years of the seventeenth century, suggests that these books were gaining in popularity and becoming steady sellers in their own right. It is reasonable to think that at least some of the audience for inexpensive books purchased a variety of offerings, including plays and books of secrets. This is supported by the ways in which printed plays were presented—with the title pages claiming that they offered the same text that had been played on London stages. The stage was remarkably accessible to a wide range of socioeconomic classes, and while the playbooks were more expensive, it is reasonable to conclude that some spectators with disposable income directed it to copies of the plays they had enjoyed as well as manuals for improving their kitchen gardens, their luck in business, and their health.

The natures and desires of book buyers are hard to pinpoint. The libraries of notable intellectuals, such as John Dee, have given us an idea of the upper end of the print market, in which *The Secrets of Albertus Magnus* really did have a place beside Aristotle and Euclid.[29] But Dee can hardly have been considered a standard reader, given his education and natural philosophical and mystical pursuits. The literary collections amassed by wealthy men and women better resemble Jaggard's catalog in their range of topics, though they tend to feature fancier editions and larger books than can be considered broadly reflective of the majority of readers.[30] These collections, taken together, suggest that inexpensive print should not be considered to have appealed only to common readers, those with minimal disposable income and little or no formal education. A better term for this kind of print would be popular, in the simplest meaning of the word. These books had a broad audience because they resonated with the interests and concerns of a large number of people, and printers like Jaggard were successful because they understood and catered to their buyers.

Albertus Magnus was a popular success—it was licensed by three printers of the first generation of London stationers, at times in each of their lives when a steady seller would ensure their success and support investments in other titles or equipment. It was also the first book to be licensed by a key figure in the second generation of printers, a man who chose to reprint it throughout his career presumably because he knew it would sell and provide him with the funds to support his business. Two important pieces can be taken away from the connections between these four men and this small book. In the cases of John King and William Copeland, the men who established the text's reputation as a steady seller in England, it was only one among several inexpensive volumes they licensed during the busiest years of their lives. It undoubtedly contributed to their solvency, while they contributed to its place in the intellectual and print landscape of sixteenth-century London. In the case of William Seres, a man whose fortune was made by a princely gift of printing privileges, *The Boke of secretes of Albertus Magnus* became the concluding volume in an experimental but profitable line of secular books that demonstrated the diversity of audience desires. William Jaggard built on the reputation established for the book by his predecessors. He used it to establish his place in the print market, to support that place throughout his career, and to lay the groundwork for a rich and diverse print catalog that placed natural philosophy and magic beside poetry and drama as equally credible and valuable textual maps for making sense of the world.

Printers Working Together

If *The Secrets of Albertus Magnus* demonstrates the host of relationships a book of secrets could have with its various printers, then the three books of secrets directed toward women demonstrate the complicated relationships printers could have with each other. In 1573 Richard Jones became the first printer to direct a book of secrets, rather than a cookery book or conduct manual, to a female audience. He was followed by Edward Allde in 1588, who produced an edition of *The Widowes treasure*.[31] In 1596 Richard Jones combined *The Treasurie of commodious Conceits, and hidden Secrets* with two housewifery manuals printed by Edward Allde and Edward White to produce *The Treasurie of hidden Secrets*, the last sixteenth-century book of secrets to be directed to female readers. This section of the chapter will trace the connections among these men and the books they produced to entice a very specific and increasingly large audience for popular print, women.

Richard Jones began his career as a freeman of the company in 1564 and in that year he set up his first shop, the Little Shop adjoining to the northwest door of St. Paul's Church. He passed through a total of six shops before 1581, with his longest stint in one place lasting for four years against St. Sepulchre outside Newgate between 1576 and 1580. In 1581 he moved to the Rose and Crown under the Holborne bridge outside Newgate, where he remained until his death in 1602. His thirty-eight years as a printer were productive, and he was among the most prodigious creators of popular print during the second half of the sixteenth century. The *Register of the Transcripts of the Stationers' Company* records that he licensed ninety-seven titles during the course of his career.[32] It seems likely that he printed considerably more than that, however, since the transcripts show no sign of either *The Treasurie of commodious Conceits, and hidden secrets* or *The Treasurie of hidden Secrets,* both of which are still extant in multiple copies notable for their good condition, and since the notes of the Stationers' Record contain punishments for Jones's habit of printing without a license.

Regardless of the exact number of books he licensed or, in the end, produced, it is abundantly clear from a review of his titles that he significantly contributed to the growth of popular print. At least half his licensed titles were ballads, and the remainder included histories and comedies, reports of natural wonders or monstrosities, news stories, travel narratives, denouncements of Catholicism and Catholics, and conduct manuals for gentlemen, ladies, and Christian families.[33] Jones is notable for the variety of texts he printed, as well as their number. His business depended upon the production of a large number of titles covering a wide array of subjects, none of which sold for very much, but which, when sold in quantity, supported a successful printing business. A survey of the titles he licensed, as well as the texts of those for which copies remain, indicates the kind of books, pamphlets, and broadsheets that drew the attention of a London crowd. It also provides some information about the characteristics of that crowd as they were perceived by a printer whose livelihood depended upon a good understanding of the tastes and preferences of his audience.

Jones stands out from his equally successful peers because he created and promoted two books of secrets for an audience that had never before been singled out as the target for this kind of literature, women. There had, of course, been books directed toward women before this, particularly those dedicated to housewifery, cookery, and appropriate feminine conduct. Richard Jones recognized early in his career the potential of female-oriented books, perhaps because he perceived the growing number of female readers and appreciated the fact that few books directed toward a single group of readers circulated only among people within that group. This is particularly true for ballads that addressed the subjects of female chastity and the feminine mandate to resist seduction. They may have been intended, at least on one level, as a warning tale for young women, but they also would have provided titillation for readers of both sexes.[34] Jones printed a significant number of books that were evidently directed toward a female audience, and he also explicitly directed a few titles to readers of both sexes. Finally, he printed a number of ballads that would have worked on multiple levels to attract male and female readers. The first group of books held a significant place in the early years of Jones's catalog. In 1570 and 1571, for example, he licensed two titles directed toward female interests, a conduct book and a ballad for maids and young wives.[35] He continued his interest in books that specified a female audience in 1573 with *The Treasurie of commodious Conceits, and hidden secrets and may be called, The Huswives Closet, of healthfull provision,* and again in 1576, with *A walkynge Ladyes nowe goo we somme pleasant things to viewe and see.*[36]

In fact, two books among his final titles included a 1595 female-directed fashion diatribe and the 1596 edition of *The Treasurie of hidden Secrets.*[37] In 1578 he printed a ballad delineating the behaviors of unchaste women that was directed toward similarly wanton readers but would likely have attracted its share of male readers interested in salacious details.[38] He continued printing for that audience in 1590 with *A warnynge for maydes to keepe their good names,* a title nearly as effective as the 1578 ballad for attracting male and female readers interested in learning to keep, lose, or encourage the loss of good reputations.[39] Jones also produced some books explicitly marketed to both male and female readers. In 1579, for example, he licensed a conduct manual for Christian youths of both sexes.[40] He seems to have realized the attraction of this kind of book, at least for readers in the early 1580s, because he quickly produced another conduct manual that contained directives for men and women.[41] His last effort in this direction came in 1590, when he produced *Articles of household discipline, conueyinge precepts and prohibicions meete for a chrystian family,* a book intended for men, women, and children.[42] During the course of his career, he printed at least ten titles that were specifically directed toward women, and very few of his other titles (probably only the male conduct books) excluded female readers. The majority of his books, perhaps all but the histories, were accessible to everyone who could read because of their length and the type of language. Jones was a master of producing popular print, and he is notable for his dedication to satisfying the needs of every type of reader rather than assuming a static, entirely masculine, population.

Jones did not print all his titles alone. In fact, his lengthy career was marked by a series of alliances with other printers with whom he shared titles and occasionally a device. The first example of this came in 1566 or 1567, when he and John Allde jointly licensed a tale of a monstrous infant.[43] John Allde was a charter member of the Stationers' Company who achieved his freedom in 1555 and printed steadily until 1584 when his son, Edward Allde, took over his business and developed it into a trade print shop.[44] Edward White, who printed several books of secrets, was a regular partner of Allde's, as well as a partner of

Jones's in 1588. Before Jones made an alliance with White, however, he teamed up with John Charlewood, another charter member of the company, to produce *A warnynge songe to Cities all to beware by Andwerps fall* in 1577.[45] In 1579 the two men teamed up again to purchase fifteen titles from Henry Denham for a total of 7 shillings, 6 pence.[46] Robert Jones's next partnership occurred in 1588 and included Edward White, half of the team that would produce *The Widowes treasure* in the same year and that would print both parts of *The Good Huswifes Jewell*, which later appeared as part of Jones' 1596 edition of *The Treasurie of hidden Secrets*. The two men, along with Thomas Orwin and Henry Carre, licensed the story of a monstrous woman who would soon be appearing for the entertainment and education of London crowds.[47] Richard Jones made his final alliance in 1594 when he joined James Roberts in licensing two titles.[48] Roberts became John Charlewood's heir after his 1593 marriage to Alice Charlewood, and it is possible that the partnership between Roberts and Jones descended from the earlier alliance between Charlewood and Jones. The fact that Richard Jones used one of Charlewood's devices for a period after his death, before passing it on to James Roberts, supports this conclusion.[49]

Two of the men who shared printing licenses with Richard Jones and one of their heirs contributed to the growing body of literature directed to female readers. In the 1580s they began to produce multiple editions of a new book of secrets for women and two new housewifery and cookery books that would later be compiled into the 1596 edition of *The Treasurie of hidden Secrets*, which was also directed to female readers. In 1587 John Wolfe printed *The Good Huswifes Jewell* for Edward White. The book was reprinted by Edward Allde for Edward White in 1596 and 1610, and it appeared in excerpted form in Richard Jones's 1596 *Treasurie*. In 1588 Edward Allde printed *The Widowes treasure* for Edward White. James Roberts printed the same book for Edward White in 1595 and 1599. In 1597 Edward Allde printed *The second parte of the Good Huswifes Jewell* for Edward White, the book that provided the last textual piece of the puzzle of *The Treasurie of hidden Secrets*. In order to understand the connections among the men who printed these books and who combined each others' productions to create new books, it is useful to examine the career trajectory and relationships of the man who had a hand in all but the first of these books of secrets for women.

Edward White began his career as a bookseller in London in 1577 and made his first entry in the registers of the Stationers' Company in that year. He was admitted to the membership of the company in 1588 and made a steady living printing and selling a variety of books, including the works of Robert

Greene, several books of secrets, and a variety of ballads and other forms of popular print. His first partnership with a printer occurred in 1578, when he teamed up with John Charlewood to produce a book containing Calvin's sermons.[50] He produced another religious book with Charlewood in 1581, then moved on to issue one book each with Robert Bourne and John Wolfe in 1591 and 1592, in addition to *The Good Huswifes Jewell*, which he printed with John Wolfe in 1587 and *The Widowes treasure* with Edward Allde in 1595 and 1599.[51] He began his partnership with Adam Islip in 1596, when the two men produced *A booke of secrets*, a small book translated from the Dutch that contains the secrets of ink making and engraving and which fits nicely into the craft tradition of books of secrets.[52] They paired up again the next year on another English translation, this time of a Spanish religious treatise allegedly compiled by Father Don Antonio de Guevara, the preacher for Charles V, king of Spain and Holy Roman Emperor.

White made a new partnership in 1600 with James Roberts, with whom he printed *The Treasurie of hidden secrets*, complete with Richard Jones's letter to the reader. This text is identical to the 1596 version by Richard Jones, and it seems likely that White borrowed, leased, or purchased the title from Jones. At least the *Transcripts of the Stationers' Register* show no sign of White paying a fine for printing the text without a license.[53] The alliance between White and Roberts, which produced another title in 1602,[54] may indicate the importance of inheritance in determining partnerships, since Roberts was John Charlewood's heir. It also supports the notion of a group of connected printers producing books of secrets, since Roberts would go on to print Gervase Markham's equestrian addition to the genre in 1595. This point is also supported by the collaboration among Edward Allde, William Jaggard, and Edward White to produce the 1607 edition of Markham's second book of equestrian secrets.[55]

Edward White's most prolific partnership was with Edward Allde, a trade printer with whom he printed and sold at least ten titles between 1588 and 1607. Of these ten, four were *The Widowes treasure, The Good Huswifes Jewell, The second parte of the Good Huswifes Jewell*, and *The Treasurie of hidden Secrets*. Two more of their titles were aimed specifically at female audiences, and a third was intended for gardeners, a population that would certainly have included women.[56] The first two, printed in 1603 and 1605, were instructional manuals that aimed to improve their readers' minds and thus their lives. The gardening book, another 1603 production, was intended to instrumentally improve readers' lives. It brought the natural world into their service in the name of producing inexpensive, wholesome vegetables to enhance the family table and satisfy hungry stomachs. These books, taken together, indicate that Allde and White appreciated that women's obligations existed in the social and natural worlds and that they would exercise their not inconsiderable buying power to better satisfy those obligations. *The Widowes treasure,* their first collaboration, their first book of secrets, and their first effort for a female audience, set the standard that both men attempted to follow in the rest of their efforts for women readers.

Printers and Authors

If Edward White's books of secrets for women exist in a tangle of his relationships with other printers, then his literary printing efforts are tied to the work of the writer and dramatist Robert Greene. Together, these two men provided London readers and playgoers with a great deal of entertainment and several characters to people the rich terrain of popular culture. Robert Greene was born in Norwich around 1558 and earned his position among the "university wits" with a bachelor's degree from St. John's College at Cambridge in 1578 and a master's degree from Clare Hall at Cambridge in 1583. He is best known for his profligate lifestyle, which included leaving his wife and son as soon as he had exhausted his marital inheritance and drinking himself to a destitute death in 1592. He is also remembered for his talent with a quill; he wrote several excellent plays to grace London stages and, according to many scholars of Elizabethan drama, influenced William Shakespeare.[57] Greene was also a poet and the author of multiple pamphlets and small books. He was famous in his own time for his tales of "cony catching," which present an elaborate world of thieves and vagrants characterized by its disdain for bourgeois morality and legality, sex, the unique grammar and vocabulary of thieves' cant, and its own social and ethical structure.[58] His reputation as a drunk and mischief maker, combined with his status as the author of these pamphlets, earned him a name as an author and a character in sixteenth-century London. His reputation was guaranteed by his infamous attack on the London theater scene, *A Groatsworth of Wit,* in which he repented his own excesses while attacking the behavior of his contemporaries, including Marlowe, with whom he had worked on several plays, and Shakespeare. His name on a frontispiece or stage bill increasingly held the power to attract an audience, a talent for which Edward White had great appreciation.

Edward White committed a relatively large percentage of his catalog to

Robert Greene. He printed at least seven Greene titles between 1587 and 1600, including histories, debates, poetry, and dramatic works. Seven titles may seem insignificant in a catalog as rich as White's, but they stand out upon closer examination for three reasons. First, Robert Greene is the only author to have had more than one of his works licensed by Edward White. Second, only ten of White's licensed and printed titles could be considered dramatic, poetic, or literary, and seven of those were by Greene.[59] Finally, White's titles represent Greene at different points in his career, as he rose from a fluent writer of fancies to a successful playwright, poet, and near-legendary scoundrel.

A review of these titles, placed in the context of Greene's authorial trajectory, reflects the place he held in White's catalog and lends some insight into the part he played in the broader market of English print. Greene's first appearance on booksellers' shelves came in 1583, when Thomas Dawson and Thomas Woodcocke printed *Mamilia: A mirrour or looking-glasse for the ladies of Englande*, a tale of the depths to which lust-ridden gentlemen will go in the quest for love.[60] In the same year, Henry Middleton and William Ponsonby printed their own version of *Mamilia* with an additional text, *The anatomie of lovers flatteries*, to protect them from accusations that they had printed an already licensed text.[61] By 1584 Greene had three books in print by different publishers, namely William Ponsonby, Roger Ward, and the triad of John Kingston, John Charlewood, and Edward White.[62] These books are noteworthy for their geographic distribution, achieved by the range of locations from which their printers sold them, and their range of genres: the romance *Gwydonis*, the religious morality tale in *The myrour*, and the collection of epigrams and entertaining debate about the nature of love in *Morando*.

Greene continued to make his mark in the second half of the 1580s so that he had ten new titles and at least two reprints on shelves across London by 1589. These books built on his solid print foundation and established him as an author who could turn his hand to topics from astrology to divine punishment with entertaining results.[63] Furthermore, they cemented the author's connection to Edward White. During the seven years between the appearance of Greene's first work and 1589, White sold four of his books, including one reprint, more than any other single bookseller. This would remain true through the beginning of the seventeenth century, with Edward White selling seven of Greene's approximately forty-three printed works.[64]

Edward White, despite being Robert Greene's most frequent bookseller, was selective about the works in which he decided to invest. Nothing represents this so well as the absence of Greene's cony-catching books from White's catalog in a period during which he clearly continued to sell the author's work. An examination of the cony books in comparison with those that White chose to sell will reflect the position he reserved for Greene's works on his shelves and, more important, the position he expected Greene to fill in the literary marketplace. It will also provide a comparison between John Wolfe, the most prolific of the cony printers, and Edward White, and the different ways in which they invested in Robert Greene. The first of Greene's exposés of the London underworld, *A notable discovery of coosenage,* was printed in 1591 by John Wolfe for Thomas Nelson, and *The Second part of conny-catching* followed in the same year, again printed by John Wolfe, this time for William Wright.[65] The next year, *The Third and Last Part of Conney-Catching* appeared from the presses of T. Scarlet for Cuthbert Burby. *A disputation between a hee conny-catcher and a shee conny-catcher* and *The defence of conny-catching* were the products of a 1592 collaboration between A. Jeffes and Thomas Gubbins, although the latter book was sold by John Busby as well.[66] The final cony-catching installment by Robert Greene, *The blacke bookes messenger,* was printed in 1592 by John Danter for Thomas Nelson.[67]

The existence of six small books containing variations on the same subject within the space of two years indicates that Greene had found a hot topic in his depiction of English criminal culture. The fact that these books also saw multiple printings, indicating that they sold sufficiently well the first time to warrant further runs, attests to the books' popularity, while the appearance of other books that built on Greene's tales of the strategies of thieves and whores indicates the heights to which he rose as a popular success.[68] Very little is known about the ways in which early modern authors marketed their texts to printers, although aberrations like Jonson, who insisted on a tremendous amount of control over his copies, have attracted some attention.[69] Robert Greene's relationships with the various people who printed his work are not well known, but it is likely that he had little control over his work once it entered the print shop. I would contend that Greene became a hot property for printers such as Edward White, who appropriated specific aspects of his image and reputation to guarantee the success of his titles. This is demonstrated by the ways in which Robert Greene appears on the frontispieces of the cony-catching books. In both 1592 editions of *A notable discovery of coosenage,* the author appears as R. Greene, after the title of the book and the characterization of its purpose as "Written for the general benefit of all Governors, Citizens, Apprentices, Countrey Farmers, and yeomen that may fall into the company of such coosening companioins" and concludes "Nascimur pro patria." Greene's last name is followed by "Maister of Arts." By the second of the cony-catching books, however, Greene's name and his academic credentials have given way to the patriotic sentiments of the first edition. The frontispiece announces the title, then the phrase "Mallem non esse quam non provesse patria" and the initials, R. G. The emphasis on the potentially nationalist benefits of the book justifies its questionable contents and makes it appealing to a broad audience, while the author's initials indicate that his name and degree is less important than his membership in the English public and represent an attempt to make him more like his readers.

This trend continues in *The third and last parte of conney-catching, A disputation betweene a hee conney-catcher and a shee conney-catcher,* and *Blacke-bookes messenger,* all of which bear the phrase "Nascimur pro Patria" followed by the author's initials. The reduction of Greene to these two letters reflects his growing popularity, as does the pun on the frontispiece of *The Defense of Coney-Catching,* which was attributed to "Cuthbert conny-catcher, Licentiate at Whittington Colledge."[70] By the end of 1592, readers associated Robert Greene with his works on the London underworld; his name was no longer necessary.

By the time Greene was established as a popular author, he had also made his name on the stage. The long, successful runs of several of his plays, both in London and in the provinces, should have made them easy sellers as printed texts, but they did not appear in print, for the most part, until after Greene's death. The period between 1590 and 1594 was instead characterized by Greene's appropriation as a warning to other educated young men. The 1590 *Greenes mourning garment: given him by repentance at the funerals of love* and the post-mortem imprint of *The Repentance of Robert Greene* emphasize his wild lifestyle and subsequent bad end. They are also likely to list his academic credentials, which uniquely qualified him to speak to young gentlemen who might be tempted into a lifestyle like his own.[71] By 1592 Greene's life and death had become sufficiently renowned that titles could entirely ignore his credentials, depending instead upon his name to bring up all the information that a reader could need.[72] *Greenes Groatsworth of Witte, bought with a million of repentance, Written before his death and published at his dying request* is an excellent example, since it reminds readers of the circumstances of his death, which would likely have prompted recollection of his lifestyle and background.

The plays, which began to appear in print around this time, presented a second image of Greene. They tended to subjugate Greene's university degree to the success his work had found on the stage. *The historie of Orlando Furioso*, for example, was printed in 1594 with the subtitle *As it was plaied before the Queenes Majestie* and no mention at all of the author, while *The honorable historie of frier Bacon and frier Bongay* appeared in the same year with the subtitle "As it was plaid by her Maiesties servants. Made by Robert Greene Master of Arts."[73] The various presentations of the dramatic quartos represent attempts to maximize their salability through an alliance between the plays' established popularity and their author's academic credentials, which might attract some readers while alienating others.

The four images of Greene described here, from man of the people to learned academic, appeared on frontispieces produced by several printers and seem to correspond best to the genres of the different texts. None of these printers was so consistent in his production of Greene as an author, nor so significant in printing his titles, as Edward White. I think a careful examination of the texts White chose to print and the authorial image he chose to present suggests an elegant understanding of Greene's reputation and position in the print market, as well as a careful manipulation of that position to maximize the printers' catalog of literary works by taking advantage of the author's already established popularity.

Edward White chose not to produce any of the cony-catching pamphlets, even while his catalog would easily have encompassed them and they are likely to have made him money. He also printed only one of Greene's plays, and he noted the author's name and degree on the frontispiece. Furthermore, the titles he sold by Greene represented the most "literary" in his catalog, poems, histories, and plays rather than wonder tales or stories of divine intervention. This tactic allowed White to make money while presenting Greene's most educated face to readers. It also allowed him to extend a clear invitation to "gentlemen readers" without refusing access to the rest of London. This approach is apparent from White's first printing of a Greene text in 1584. That year, John Kingston, John Charlewood, and Edward White printed and sold *Morando the tritameron of love* and attributed it to "Robert Greene, Maister of Arts in Cambridge."[74] By 1587, when John Wolfe and Edward White printed *Eupheus*, White had further developed his strategy for selling Greene's books. This book is not only attributed to Robert Greene, but it gives the attribution and the author's title in Latin and is directly addressed to the needs of gentlemen, boasting that it contains the "vertues necessary to be incident in every gentlemen."[75]

White's next edition of a Greene book depends more upon the dedicatory epistle and the letter to the readers to appeal to gentlemen and, unlike all the others, fails to cite the author on the frontispiece. *Perimedes the Blacke-smith* has something to offer a wide variety of readers, since it contains "special principles fit for the highest to imitate and the meanest to put to practice." This invitation to the peerage and peasant alike fizzles at the last line of the frontispiece, which only the educated could have grasped: "*Omne tulit punctum, qui miscuit utile dulci.*" It founders entirely in the dedicatory epistle, which is aimed specifically at securing the support of Gervase Clifton, Esq., for the author and contains a host of easily missed classical allusions. That failure to address a broad variety of readers is also represented in the letter to the readers, which begins "To the Gentlemen readers."

Any doubt that this book found its widest audience among the educated is entirely dispelled by the first of its "pleasant exercises," a poem by Greene in French.[76] By 1592 White had lost his brief penchant to advertise Greene's books to the masses. His edition of *Philomela, the Lady Fitzwaters Nightingale* is notable for its brief title and the authorial attribution to "Robert Greene Utriusque Academiae in Artibus magister." It is also notable for the dedicatory epistle, which begs Greene's case with another aristocrat, Lord Fitzwater, and the letter "To the gentlemen readers," which shares the heading but not the content of the 1588 letter.[77] The last text by Robert Greene that Edward White printed during the sixteenth century represents the successful realization of the multiple available images of Greene coalescing into a recognizable and highly salable name. White ensures, however, that the name Greene is never far from its academic credentials; in his edition of *Greenes Orpharion,* "Robertus Greene, in Artibus Magister" is the last line before the printing information and reinforces the title. The book is dedicated to Robert Carey, Esq., and the letter to the reader again addresses the audience as gentlemen. It is, however, unique among Greene's letters in blaming the printer for the delay in the text's production. A perusal of the letter would suggest that Greene was indeed alive and charmingly apologetic about the long time his readers might have waited for his *Orpharion,* rather than having died seven years before without ever seeing his book in print. White was the only printer to embrace Robert Greene's literary texts when he was best known for seamy tales of the English underworld, and the letter reminds us of how carefully he managed the author's postmortem literary career by capitalizing on his popular print and stage successes and encouraging his slowly building reputation in English literature.

Edward White not only contributed to Robert Greene's career, he also shaped Greene's lasting gift to dramatic popular culture, the characters Friar

Bacon and Friar Bongay. He printed *The honorable historie of frier Bacon and frier Bongay* in 1594, when Greene was available to the public in all four of his characterizations and thus very visible. The play appeared with Robert Greene's name displayed prominently below the name of the company that had performed it most recently in London, and his name was followed by his academic credentials. According to Henslowe's diary, this play was already considered old by 1589 when it was performed by Lord Strange's men, and it seems to have played multiple times in London as well as in the countryside. The story had broad appeal, since its primary characters ran the gamut from the king of England to a pretty milkmaid, and none of them was treated so harshly that audience members or readers would have been alienated. The themes were also familiar and accessible, stressing the importance of obeying the limits of natural knowledge and subjecting the quest for knowledge to the needs of the nation. Finally, the play's settings would have had broad appeal: It is based partially in the pastoral landscape of the English countryside and partially in the academic enclave of Oxford University and includes some nice moments of natural magic to improve both locations.

The play is charming, its characters are appealing, and it speaks to the nationalist and imperial concerns of the early 1590s. Edward White helped it to move from its position as a successful theatrical trifle to the world of print with his 1594 edition, and in so doing, he spread the characters of Friar Bacon and Friar Bongay, Greene's masters of natural magic, to a broader audience. Some members of that audience were most certainly writers themselves, and at least one produced a chapbook detailing the further adventures of the frantic friar and his partner that printed well into the seventeenth century. The figure of the academic magician was not, of course, invented by Robert Greene; he simply invented one who was funnier and more broadly appealing than any of the other tortured souls who conjured devils on London stages during that period. In fact, he invented characteristics for a historical figure already appreciated for his talent for manipulating unseen forces, and he did it well enough to ensure that Roger Bacon would have his own book of secrets in the London bookstalls by 1597. Friar Bacon paid him back a thousand times, as we will see in chapter 3; since his legend kept Greene's name alive long after he might have fallen silent among the babble of sixteenth-century playwrights and the academic fascination with William Shakespeare.

Conclusion

This chapter has traced a number of the ways in which the developing print marketplace, the community of printers, and the growing population of authors combined to produce various kinds of books of secrets and, in one case, a literary reputation that produced its own book of secrets in sixteenth-century England. These examples demonstrate a lively and highly interactive print culture in which printers competed and cooperated and carefully planned the books they would print, the ways in which they would produce them, and the audiences who would have access to them. I have introduced three sets of relationships that seem to have been central to the creation of the genre of books of secrets and, in particular, books of secrets aimed at audiences characterized by readers' sex, educational level, and socioeconomic status.

First, I traced the relationship between four printers and a single book, the highly durable *Secrets of Albertus Magnus,* which played a very different role for the three men who printed it during the early years of the Stationers' Company and the single member of the second generation of organized London printers who introduced the book to the seventeenth century. Second, I traced the relationships among the printers who created the only three books of secrets printed specifically for women during this period. The fact that these books were so clearly directed to female readers and contained a set of rules governing female manipulation of the natural world, as will be developed in chapter 4, indicates that printers appreciated the buying power and interests of female readers. These books are also remarkable because they stretch our ideas about the kind of work women were doing and the kind of equipment to which they had access, as well as demonstrating the existence of books for women concentrating on general natural knowledge, rather than simply female and pediatric health care or cookery and sewing. Finally, they stand out because they represent three decades of interactions between printers whose ties range from inheritance to trading copies to a steady partnership.

Third, I emphasize the importance of alliances in producing books of secrets. The permutations of the early modern print market remain hazy, and an examination of the relationship between printers of books of secrets demonstrates that connections created in the first generation of the Stationers' Company had lasting effects on the ways in which printers interacted, who partnered with whom, and who sold what throughout the sixteenth century. Finally, I explored the relationship between an author notable for his productivity, popularity, and short career and his printers. Robert Greene did not write a book of secrets, instead he created a character who, after being made available to the public through a quarto edition by Greene's most faithful printer two years after his death, took on a life of his own and soon had his own book of secrets. Greene's career is fascinating because of the wide variety of books he wrote and the ways in which printers made use of his name and academic credentials to produce authorial characterizations to suit each of his genres. I have no evidence that Edward White ever met Robert Greene, but I don't believe that this argument depends on physical contact—instead, it relies upon the value that White put on developing a literary aspect to his catalog and the way he developed Robert Greene's persona to suit his ideal of a man of letters.

The introduction to this chapter promised to place books of secrets within the print marketplace. It demonstrates that books of secrets and books dedicated to the explication of natural knowledge, especially those that promised readers a more manageable or predictable world, formed a recognizable genre within print culture. Furthermore, it suggests that these books were the product of a group of men even more tightly connected to each other than they were to their colleagues. While every printer in London was required to be a member of the Stationers' Company, the men who produced books of secrets were also bound by ties of blood, apprenticeship, and partnership that are not characteristic of the printers of dramatic quartos or as neatly predictable as the connections between the men who printed almanacs, whose rights were often guaranteed by royal privilege. I lack the evidence to contend that these men were somehow motivated by a shared interest in natural philosophy or natural knowledge, but I do think that some explanation of their printing habits is possible.

David Hall[78] was among the first to contend that almanacs and books of knowledge were steady sellers because they presented systems for predicting nature, including disease and the weather. I would argue that books of secrets were appealing to readers for a similar, although more complicated, set of reasons. The first thing to note is that, once printed, secrets were still secret—each book contains a testimony to the ways in which the collector of the presented knowledge struggled to find the information in the book and agreed to print it only under great duress. They each also argue that the knowledge they present is usable only by a diligent, very careful reader who can follow the entire text rather than attempting to make use of it for a single recipe that would permit personal gain. Second, if printed secrets are still secret and mastering them represents entry into an exclusive group of clever readers with the ability to manipulate the natural world, then it is no surprise that books of secrets sold extremely well.

Finally, the books of secrets I discuss are directed toward readers who would not otherwise have had access to the printed promise of control over nature. Women, servants, and others who lacked university training, particularly knowledge of Latin, would not have been able to read the plethora of ancient and academic texts that promised access to the world's secrets. They did, of course, have their own systems for understanding and making predictable the natural world, but the promise of an entrée into formerly reserved knowledge would have been hard to resist. It would have become harder to resist in the face of affordable prices and letters to the readers suggesting that the text had been compiled by someone like them, if blessed with the privilege of more time and the funds to chase down the secrets of the natural world. The printers who produced these books were engaged first and foremost in a struggle to survive the competitive print marketplace and make a profit, and books of secrets certainly looked as though they would contribute to that goal. They did so by inviting readers who had previously been ignored to spend their three or four pence on a book that would return their investment in the shape of gold, better health, a better laid table, a better run stable, and, most important, a sense that they had some influence over their world.

CHAPTER TWO

Roger Bacon, Robert Greene's Friar Bacon, and the Secrets of Art and Nature

> "For if thy cunning worke these myracles,
> England and Europe shall admire thy fame,
> And Oxford shall in characters of brasse,
> And statues, such as were built up in Rome,
> Eternize Friar Bacon for his art."[1]

To this point, I've concentrated on the print marketplace and offered evidence that there was a small group of interconnected printers producing and selling books of secrets. In this chapter, I will develop the role of authors, or more accurately the creation of authority, in cheap print. I will examine the three kinds of external authorities that appear as experts on the title pages and within the contents of these books, and I will link those iterations of authority to broader social and cultural movements within intellectual trends in sixteenth-century England. Finally, I will proceed to examine the collaboration of various aspects of popular culture in appropriating, reconstructing, and promoting a particular master of secrets: Roger Bacon.

From *The Erra Pater, or the Book of Knowledge* to *Aristotle's Masterpiece* and *The Book of Secrets of Albertus Magnus*, the English cheap-print market is full of small books attributed to commanding authorities. The first thing to realize is that none of these men had anything to do with the books they were supposed to have written, and some of these laudable authorities were not even real. In order to understand the ways in which authority was constructed and used in early modern print culture, it is useful to break down the types of people given authorial status as masters of the natural world. The first type consists of well-established classical figures such as Aristotle and Pliny, whose works had dominated academic culture for so long that their names and reputations had percolated out of universities and into common knowledge. The second type consists of mythical figures such as Erra Pater and Hermes Trismegistus, whose romantic stories and ties to cultures venerated for their mastery over nature established their authorial potential. The final type includes figures such as Albertus Magnus and Roger Bacon, medieval natural philosophers and theologians who had devoted their lives to pursuing the secrets of nature and the potential for art to manipulate the natural world. The available remnants of their legitimate works attracted the attention and respect of early modern natural philosophers, while their names and interests became synonymous with natural knowledge in the rapidly changing domain of popular natural print.

Veneration of the classical world was a hallmark of the Renaissance, and it played an important role in shaping sixteenth-century English literature and intellectual culture. Humanism made a relatively late entrance into English culture, but as it gained court support, its emphasis on obtaining original-language copies of foundational texts such as Aristotle's works and writing new ones in their image quickly spread. The English Renaissance was not, however, a copy of the Italian one that preceded it, and as Richard Helgerson has established, it produced literature that combined medieval and classical ideologies, rather than rejecting the former in favor of the latter.[2] Books of secrets accomplish something very similar, since their contents combine folk knowledge, largely composed of general systems of sympathy and antipathy, and classical ideas about the natural world and their title pages frequently attribute those contents to great classical or medieval magi. William Eamon has argued that this combination of legitimate scholarship and folk wisdom could have threatened the legitimacy of books of secrets and undermined their place in the circle of academic natural philosophy.[3] I would argue instead that it makes them particularly useful as a bridge between multiple worlds, academic and popular, classical and medieval, "real scholarship" and folk knowledge. That bridge calls into the question the strict divides many modern scholars believe to have existed between those worlds and suggests a much freer and more flexible marketplace of ideas.

Nothing demonstrates the mixture of high and low culture better than *Aristotle's Masterpiece*, a combination of folk remedies and humoral explanations for the problems that accompany menstruation, menopause, birth, and early childhood. Aristotle himself, of course, paid very little attention to questions of gestation and birth beyond the question of how to predetermine a baby's sex, and so the information presented in this book cannot have come from his corpus, which still formed the foundation of medical school reading lists during the early modern period. Because of his place within academic medical and ancient Greek culture, Aristotle was associated with humoral theory, the system that explained disease and health according to balance among the humors in the body. Humoral theory was also part of vernacular ideas about health and disease, which meant that it formed part of the common ground between elite and folk knowledge about the body and its relation to the natural world. The small medical book attributed to Aristotle functioned in a similar way, and it contained information that would have been useful and accessible to medical students and midwives, as well as women interested in taking care of their own and their children's health and young men seeking sexual information and titillation.[4]

All the books of secrets that will be analyzed in this study include references to Pliny, Aristotle, and Hippocrates and other men whose names were synonymous in academic and popular culture with expertise about and mastery over the natural world. By attributing information to experts such as these, the books gained an aura of legitimacy and were linked to a long tradition of natural philosophy that had significant cultural resonance. They also supported the importance of witnessing and experience in making claims about the systems influencing the natural world. As Steven Shapin[5] has demonstrated, seventeenth-century gentlemen natural philosophers increasingly embraced an approach to the study of nature that emphasized their status as witnesses to the phenomena they were reporting. The story that Shapin tells about the place of witnessing in elite laboratory closets has relevance for a much broader audience than the one for which he intended it. Books of secrets also emphasize witnessing, both by using classical authorities as witnesses to lend credence to the information they contain and by privileging readers' experiences with the natural world as legitimate and important. The latter strategy will be explored in the next chapter, but the importance of well-known names on title pages and in the contents of popular natural philosophical texts would have served a significant purpose in tying them to the academic tradition of natural philosophy and inviting readers who did not have access to that world to pay a small sum for entrée into the arena of natural secrets.

The classical world was not the only one to supply these small books with recognizable authorities to testify to the systems governing the natural world and the phenomena they produced. The wandering Jew "Erra Pater" was the creation of a sixteenth-century author or printer, and he became the "author" of a longstanding steady seller.[6] His association with the Jewish faith, which was noted for its study of the Kabbalah and ensuing mastery over nature, lent it authority and legitimacy, as well as an enticing air of exoticism that would have made it attractive to readers from a variety of intellectual backgrounds. According to late medieval lore, Hermes Trismegistus was an ancient Egyptian alchemist who used prayer and tremendous self-discipline to finally succeed in making the philosopher's stone. The method for doing so was recorded in the scroll clutched in his dead hands and transported back to Europe for translation and quick publication. Trismegistus was, instead, the creation of the late medieval imagination, but his exotic background and fabulous story undoubtedly helped to sell alchemical and natural philosophical books that contained the treatise attributed to him.[7]

Medieval scholars also had their role in selling information about nature to popular audiences. Albertus Magnus, a thirteenth-century Dominican, had an extensive reputation for exploring the power of art over nature. His name remained synonymous with natural philosophy and attempts to control the natural world in the early modern period. English scholars such as John Dee read what they could find of his work, and they also read the book of secrets that was attributed to him. That book of secrets introduced Albertus Magnus to a broader audience, taking advantage of his association with academic natural philosophy and his belief in the support of theology by natural philosophy to sell a book that provided a great deal of information about how to manipulate and take advantage of the natural world. While the contents of that book cannot be the product of Albert's efforts, his name on the title page and the connection it implied to the world of scholastic and godly natural philosophy ensured that it would become a valued asset in elite and popular print culture, as well as an addition to the growing number of texts that asserted the power of art over nature as a useful tool rather than an affront to God. Roger Bacon, a thirteenth-century natural philosopher, the star of a sixteenth-century play, and the author of a sixteenth-century book of secrets, is an interesting combination of an established natural philosopher whose reputation was extended into the popular arena by a playwright exploring the limits of the acceptable pursuit of art over nature and then adopted as a champion of art controlling nature in the guise of an author of a late-sixteenth-century book of secrets.

Roger Bacon

Little is know about the life of the real Roger Bacon. In the introduction to his critical edition of Bacon's *De multiplicatione specierum* and *De speculis comburentibus,* David Lindberg surveys the available evidence to conclude that his subject was born around 1220 as the son of a relatively wealthy family. He was educated at both Oxford and Paris, earning his M.A. at one of these institutions around 1240. During the 1240s, Bacon lectured in the faculty of arts at Paris, focusing primarily on the Aristotelian corpus.[8] During this period, he discovered the pseudo-Aristotelian *Secretum secretorum* and produced a critical edition of this text. The occult content of this book may well have interested him and influenced his research interests, but Lindberg argues convincingly that the trajectory of Bacon's later research interests suggests that it probably did not define his later interests.[9] He devoted the next two decades to developing natural philosophical, occult, and mathematical theories that derived much more clearly from Grosseteste's philosophy than from Aristotle's.[10] He did not pursue these studies as a member of the Oxford faculty, and it seems that he supported himself and his investigations with family funds.[11]

He had to forsake those funds, which had probably dwindled as a result of poor political luck, in order to join the Franciscan Order, a move he made around 1257.[12] While Bacon left no testament to his specific motivation, Stewart Easton argues that Bacon believed in a close relationship between morality and true knowledge and that joining the Franciscans might have been part of his quest for morality.[13] He could have been drawn to the Franciscans because of their reputation for supporting scholarship or because of his admiration for Grosseteste, whose work in experimental mathematics and natural philosophy had influenced his thought. The decision might have been rooted in practical interests as well, since his family could no longer sponsor his investigations, and joining a religious order would have provided an opportunity for scholarship without the demands of teaching.

Regardless of the specific combination of motivations, upon joining the Franciscans he continued his investigations of the natural world and directed them toward the question of reforming the Christian world.[14] These investigations required funding, and they led him to pursue the patronage of Cardinal Guy de Folques, the man who would become Pope Clement IV. He never received any financial support from Folques as either cardinal or pope, but he did receive a mandate to secretly produce a work on the possible remedies to be found in natural or occult philosophy and mathematics for the problems of Christianity. The relationship and its accompanying mandate put Bacon at odds with the philosophies and governing policies of elite members of the Franciscan Order at the same time that the Order was responding to a period of philosophical division with a set of new regulations about the possession and authorship of unauthorized books.

In 1260 the Franciscan order in the shape of Bonaventura and the Chapter of Narbonne responded to the furor over the publication of Gerard of San Borgo's *Introduction to the Eternal Gospel,* an ardently apocalyptic text, by forbidding the possession or authorship of any book without the permission of the minister-general. It also forbade all communication between members of the order and the pope or his representatives.[15] Bacon was in a difficult position—he had disobeyed the latter regulation by soliciting the pope's support for his investigations and was rewarded with an unfunded mandate for the secret production of a gigantic and now illicit work. The content of that work complicated his problems, since he accused the most prominent Franciscan theologians of ignorance, opposed the astrological nondeterminism of the minister-general of the order by asserting that astrology possessed some level of predictive power and was therefore a useful tool for the church, and justified his arguments in the name of defending Christianity in the impending war against the Antichrist.[16] Despite all these issues and his complaints to the pope about poor treatment at the hands of his brethren, Bacon was able to produce, partially through compilation of older works and partially through a decision to summarize his grand project, at least three major works for Clement IV.

The *Opus majus* and *Opus minus* and at least one independent treatise, *De multiplicatione specierum,* were carried to the pope by one of Bacon's students in either late 1267 or early 1268. The pope certainly received the books, but no record remains of his response.[17] The years after 1268 were clearly productive ones for Bacon, but his activities aside from writing remain very shadowy. He wrote at least four more treatises on natural philosophy during the early 1270s and continued to write until the end of his life, leaving behind an unfinished theological treatise begun in 1292. He did not entirely escape the wrath of the Franciscans, whose theology he mocked and whose rules he unmercifully broke. He may have been imprisoned in the late 1270s; his tendency toward rebellion and his radical philosophy make it seem probable, though no clear record of the charges or any sentence remains. He was certainly in Paris, probably to be kept under closer watch, for some of this period.[18] He died in Oxford, still pursuing the secrets of nature, around 1292.

Greene's Frantic Friar

By the sixteenth century, Bacon's status as a master of natural philosophy was ensured by the place of his works on the bookshelves of Englishmen obsessed with the search for control over the natural world. Evidence indicates that John Dee had copies of his manuscripts as early as the 1550s, and other educated practitioners of natural and occult philosophy considered his work important.[19] He had also begun to enter these men's imagination as a mythical master of the natural world whose desire for control over nature had driven him to exceed the grasp of his frail human capacity. The story of the brass head was probably already circulating in both academic and popular circles at this time, though it had not yet been printed in association with Bacon. The tale of Bacon's failed attempt to create a brass head that would channel preternatural forces and answer all his questions about the natural world would only have added to Bacon's appeal to Elizabethan natural philosophers bent on stretching the limits of art's control over nature. Bacon would become a force in the popular imagination as well when he caught the eye of sixteenth-century playwright Robert Greene (1558–1592). He would turn the story of the brass head into a Faustian morality tale with a twist, in which meddling with occult forces is wrong unless it is at the command of the king.

The Honorable Historie of frier Bacon and frier Bongay was printed in London for Edward White in 1594, with the note that the written version matched the one played by Her Majesty's Servants.[20] The play was performed by Lord Strange's Men in March 1591, April and May 1592, twice in January 1593, and once in July 1593. It was also performed by the combined troupes of the Queen's Men and Lord Sussex's Men twice in April 1593.[21] It probably played outside London as well, though no records remain of any provincial performances. The scholars who have concentrated specifically on this play have concluded that the printed version released by White is a relatively good quarto, or "a good text that was marked up to some extent for production."[22]

The brass head holds a significant place in Greene's play, and as Bacon's most renowned effort it has posed some difficult questions for historians of Bacon and literary scholars interested in sixteenth-century popular drama and cheap print. The vast majority of the former group has relegated the tale to the realm of popular myth and gone on to reckon with Bacon's better-documented and more evidently scientific interests, particularly his three opera. They point out, quite rightly, that this tale seems to have gained prominence approximately three hundred years after the death of its protagonist. Print historians and literary critics, on the other hand, have taken both the dramatic and chapbook renditions of the story of the brass head for granted, labeling it part of "an accumulation of legends that began to accumulate early around the historical Roger Bacon" and Bacon himself a "legendary magician and folk hero."[23] J. M. Brown asserts "around the names of Faustus and Roger Bacon had clotted all the strange legends of supernatural power. Marlowe claimed one and Greene the other."[24] The problem that remains is why Roger Bacon became the site around which legends of occult skill, nationalist sentiment, and moral quandary accumulated, or more aptly, how the myth of Bacon became the star of a moral tale about the fate awaiting those who pursue the secrets of nature for the wrong reasons and also a popular natural philosophical authority known for trying to stretch the limits of art.

Greene has attracted scholarly attention for his role as a predecessor of Shakespeare, for the plays he contributed to the English stage, for his numerous pamphlets, and for the print war over his character and talent that followed his death.[25] He was known for his derivations, and most literary critics have contended that dramatic inspiration for this play came from Lyly's *Campaspe* and possibly Marlowe's *Faustus*.[26] Some Greene scholars, however, believe that an anonymous chapbook entitled *The Famous historie of Frier Bacon. Containing the Wonderfull Things that He did in His Life: Also the Manner of His Death; With the Lives and Deaths of the two Coniurers Bungye and Vandermast* introduced Greene to Bacon and was a primary source for the play. This proposition lacks historical evidence, since there is no record of this text in the Stationer's Registry before 1624, when it saw a great deal of success and was consistently reprinted.[27] It seems likely that Greene built on Bacon's existing reputation in elite natural philosophy and at court to produce the play. Its performances and print sales, in combination with the growing Elizabethan enthusiasm for natural philosophy, especially alchemy, prompted the printing in 1597 of the pseudo-Bacon text, *The Mirror of Alchimy*, and the creation of the chapbook.[28]

Paul Dean names Greene and *The honorable historie of frier Bacon and frier Bongay* as part of the romance or pseudo-history tradition that formed the majority of English-language history plays in the 1590s and laid the groundwork for Shakespeare's *Henry VI*.[29] Douglas Petersen associates the depictions of princes in Lyly's *Campaspe*, Greene's *Honorable historie of frier Bacon and frier Bongay*, and Shakespeare's *Henry IV, Part 1*, arguing that each features a young man destined for national and historical greatness whose future is threatened by his fondness for play.[30] Kurt Tetzeli von Rosador makes an argument for similar uses of religion and magic in John Bale's *Lawes of Nature, Moses, and Christ, Corrupted by the Sodomytes*, Greene's *Honorable historie of frier Bacon and frier Bongay*, and Shakespeare's *Henry IV*, asserting that the interdependence of church and state and both institutions' efforts to contain or destroy rivals like magic is featured in all three.[31] All of which simply goes to show that Greene was a part of the "crib and share the lot" atmosphere of Elizabethan theater, right down to the 1602 entry in Henslowe's diary that renders moot any argument for sole authorship of the play: "Lent unto Thomas downton the 14 of desembr 1602 to paye unto mr mydelton for a prologe and epelloge for the playe of bacon for the corte the some of vs."[32]

There appears to have been little or no censoring of Greene's play, so the play text reflects the majority of the words that theatergoers would have heard spoken on the stage in the 1590s. The play's integrity is not surprising, given that it supported the popular notions of a strong monarch, a defended England, and a corrupt Catholic church while offering two love stories and more than a little magic. It is strengthened by the moral tale running through the plot, which cautions that men who seek to master nature for their own ends will suffer for their hubris, while those who do so for the glory of the state will be safe. The plot is relatively simple and centers on Prince Edward's desire to seduce a country girl named Margret. He fails to win her with his wily charms and turns, at the advice of his fool, to Friar Bacon for help, while sending his friend to assess his standing with the young woman. His friend, Lacy, falls in love with her himself, and while the prince watches through Bacon's crystal ball, Lacy promises to marry her.

At the same time, the king welcomes a retinue of royal visitors, including the scholar Vandermast, the king of Germany, and his daughter Elinor, who has come to marry the prince. Upon finding that Edward has gone to Oxford to meet with Bacon, they journey there, encountering Friars Bongay and Bacon on the way.[33] Vandermast challenges Bongay to a conjuring duel and defeats him, only to be defeated and banished by Bacon, whose victory is claimed by the king as a triumph for England. Bacon returns to Oxford, where his seven years of laboring over the brass head come to a disastrous end because his student fails to waken him when it speaks. He sinks into a despair that is only deepened by his role in the deaths of two young men who kill each other while watching their fathers duel to the death in his crystal ball. He swears off magic, vowing to spend the rest of his life in prayer, until the king calls him forth to prophesy in the final scene, a double wedding in which Edward married Elinor and Lacy married Margret.

The play opens with the earl of Lincoln, the earl of Sussex, a gentleman, and the king's fool attempting to discover why Prince Edward is so forlorn after a successful trip of hunting and frolicking in Fresingfield. The fool Raphe, led by his companions, asks Edward whether he has fallen in love with the keeper's daughter and offers a cure for his lovesickness. His cure, that he and Edward shall exchange dress in order to fool love, is the first appearance of deception through costume and the attendant supposition that character traits will be exchanged along with clothing.[34] Raphe justifies his plan to Edward, saying, "Why so thou shalt beguile Loue, for Loue is such a proud scab that he will neuer meddle with fooles nor children," as though by dressing the prince as a fool, he will become one and lose the traits that accompany his station and that render him susceptible to love.[35]

Edward and Raphe spend much of the their time onstage in each other's clothes, and their costume play contributes to the construction of their characters and creates opportunities that would have been impossible without the dress exchange. Dress, for example, is the reward for Raphe's plan to capture Margret for the prince. The promise of a new coat motivates his suggestion that the whole party ride to Oxford and ask the help of Friar Bacon in winning Margret's virtue for Edward, a task the latter deems impossible without the help of a magician or wedding vows. The emphasis on clothing switches also reflects the importance of systems of sympathy and their vulnerability to manipulation in this image of the natural world. The characters hope that by wearing each other's clothes, they will also take on each other's characteristics and social position. This particular example of systems of sympathy at work would not have been unfamiliar to readers of natural secrets, who would have encountered a similar example in the effects of harlots' clothing, or even their mirrors, on chaste women.[36] The fool, for example, suggests that, to prevent the prince from being missed at court, he should dress as Edward ("prince it out"), while the friar transforms Edward into different pieces of Margret's clothing so he can be close to her body.

The desirability of Margret's body, both for itself and the social and political power it represents, is a driving force in the play. Edward's fateful meeting with her is marked by its emphasis of her physicality, which is sexual in its performance of domestic labor. Edward argues that none of his companions could have appreciated Margret's full beauty because they did not see her alone, as he did when he followed her into the milkhouse and saw her "secret bewties" revealed.[37] While there is a slightly racy tone to the image of Edward following Margret into the isolated milkhouse, these are not the "secret bewties" of a naked body. Instead, they seem to have two aspects: the beauty of a woman doing productive domestic work and that of a virtuous woman who will have no part of love without marriage. Edward's decision to search out Friar Bacon for help in winning Margret's favor depends upon a veiled and failed attempt at seduction, which probably took place among the vats of cream and made him see "That marriage or no market with the mayd."[38] The secrets he wants revealed are, in the hands of Bacon, made parallel to the secrets of nature in a fascinating exploration of the ways in which the word could function in English popular culture.

Bacon enters the play as a means of accomplishing Edward's desires through magical ends, and the fool, who suggests his services, advertises him as a scholar known to be "a braue Nigromancer" talented at transforming bad or troublesome things into useful ones—"he can make women of deuils, and hee can iuggle cats into Costermongers."[39] In the fool's introduction, Bacon is a figure whose knowledge of magic and skill at manipulating the occult is for sale, even for malicious ends.[40] It is not surprising that Bacon's relationship to his art becomes more complicated when he steps onstage. Bacon appears in his cell at Brazennose College at Oxford, surrounded by his student Miles and three Oxford doctors who have come to investigate the rumors of Bacon's occult skill. These scholars establish how Bacon's name came to be on the fool's lips: "I tell thee Bacon, Oxford makes report, Nay England, and the court of Henrie saies, Thart making of a brazen head by art, Which shall unfold strange doubts and Aphorismes, And read a lecture in Philosophie."[41]

The scholars who come to visit Bacon reflect his complicated position in the academic community as either a master of the natural world and the preternatural forces that affect it, or a man whose reach has exceeded his grasp. That position would have been familiar to sixteenth-century natural philosophers, whose forays into liminal arts such as necromancy and alchemy raised eyebrows in the church and among their peers. The struggle to bend nature to human will by summoning the help of preternatural forces was not broadly accepted, and scholars who sought the aid of angels and demons did so at their own risk.[42] Bacon's proud boast that he has the moon, thunder, and the devil himself under his command leaves him open to accusations of being a charlatan or a heretic, and given the array of accusations leveled at him by his peers in the scene and the similar reactions to the work of scholars like Dee, it is hard to tell which is worse. He announces, "What art can worke, the frolicke frier knows," but his pet project, the creation of a brass head that will reveal the secrets he has long tried to discover, stretches the limits even of his deepest conception of art.[43] He justifies his desire for control over the natural world and his willful plans to stretch the limits of art with the potential his project has to strengthen and defend England, a godly aim that rescues the brass head from the ignominy awaiting any product of a charlatan or an apostate.[44]

His colleagues agree that natural philosophy does not stretch far enough to accomplish Bacon's grand plan, but they remain divided over the propriety of engaging with preternatural forces to make it do so. One urges the value of magic in such a case, advertising its potential to manipulate hidden forces left untouched by natural philosophy, "No doubt but magicke may doe much in this, For he that reades but Mathematicke rules, Shall finde conclusions that auaile to worke, Wonders that passe the common sense of men."[45] This scholar believes in both the existence of such wonders and the potential of magical knowledge systems to access and manipulate them, but it is not clear exactly what he means by magic. Since he is speaking in reference to Bacon's necromancy, it seems likely that here magic means meddling with preternatural forces in order to bend the natural world. His more skeptical friend argues that talking brass heads go beyond art and magic into the realm of hoaxes and fairy tales. "Haue I not past as farre instate of schooles: And red of many secrets, yet to thinke, That heads of Brasse can utter any voice, Or more to tell of deepe philosophie, This is a fable that Aesop had forgot."[46] He accuses Bacon of trying to earn fame through trickery, reducing his claims of mastery over occult forces to a set of parlor tricks played by a charlatan.

This argument would have been familiar to early modern natural philosophers who were struggling to discover the extent to which art could influence nature without losing its base in the natural and sliding into the uneasy and increasingly illegitimate realm of magic. The men in the play, like late-sixteenth-century academic natural philosophers, are debating more than Bacon's skill; they are questioning whether human knowledge is limited by God or by the limits of art, and they are questioning the characterization of a man who investigates and stretches those limits as a master scholar or a trickster. Bacon draws his doubting colleague into an exchange that shows off his skills, pokes fun at the scholar committed to the narrowest possible definition of art, and elaborates on the complicated web of associations, already introduced by Edward, among women, nature, books, and secrets. The exchange will end in Bacon's triumph among his peers.

The display begins with Bacon asking his doubting colleague whether he had spent the previous night at Henley-upon-Thames. The man agrees that he had, but when asked what book he studied there, he claims to have studied nothing. Bacon promises to reveal the book through conjuring: "Ile shew you why he haunts to Henly oft, Not doctors for to tast the fragrant aire: but there to spend the night in Alcumie, To multiplie with secret spels of art."[47] He conjures a demon to do his bidding and produces not a book, but the hostess of the inn at Henley, the colleague's hidden text. Working women, particularly those engaged in domestic productive tasks, are full of secrets, as Edward noted in his description of Margret in the dairy.[48] Bacon complicates that relationship by inscribing those secrets in the "book" he conjures from Henley.

For these men, books are filled with secrets that can be learned and appropriated through careful study, and the metaphor of the woman as book implies that she can also be mastered through devoted attention and the application of art. The art Bacon chooses to extend is alchemy, whose alliance with metaphors of generation and reproduction is embedded in its theory.[49] Alchemy is a highly gendered discipline, with masculine and feminine traits assigned to the metals being acted upon, the alchemist, and nature itself. A completed alchemical recipe depends upon the marriage of metals whose gender affiliations frequently switch, so that the liminal figure of the hermaphrodite mercury becomes the primary mover and the most useful sexual model.[50] It is frequently associated with sexual reproduction, so Bacon is making a lewd joke at his colleague's expense—that he goes to Henley to have sex with a bar wench. The joke has another level, however, in which women and books are allied as the objects of male interest and desire. While books contain contents that can be read and mastered, women and nature hide their secrets and are reluctant to give them up to any man unwilling to stretch himself and his art to achieve his goals. Bacon's approach to natural philosophy and romance triumphs here, and he becomes the right man to seduce a milkmaid for the prince and summon the help of demons to protect England and reveal the secrets of nature. His student Miles confirms his victory with another lewd joke: "Ile warrant you maister, if maister Burden could coniure as well as you, hee would haue his booke euerie night from Henly to study on at Oxford."[51]

Bacon's authority within England is established by this debate, but it remains to be proven against other great scholars of the age. Academic competition recurs in the play and gives Bacon and England the chance to match wits with a great German scholar and the nation that produced him. The king of England has welcomed a visit from his son's proposed bride Elinor, and she is accompanied by a party that includes Vandermast, a German academic who came to England specifically to see if he could outsmart his peers at Oxford, having already defeated the best scholars of many of the greatest European universities. King Henry welcomes Vandermast but warns him that he will face a tough match in Oxford, claiming friar Bacon's skill as "England's only flower" and a credit to the nation's prowess in the academic arena.[52] It is odd, after the king's praise of Bacon, that he is not the first scholar to face the visitor. Friar Bongay gets that honor, and he begins by conjuring the tree from the garden of Hesperides, complete with golden leaves and a dragon shooting bursts of flame. Vandermast counters by conjuring a demon-Hercules to break down the branches of the tree. All looks lost when Bongay cannot stop the demon from destroying the tree and capturing the golden fruit. Vandermast deems him "learned enough to be a Frier" but no match for himself and calls upon Henry to crown him with laurel as the kings of other failed scholars have done.[53] Bacon appears before that can happen, and his status as the king of the preternatural realm is established by a speech from Vandermast's Hercules-demon. The demon refuses to obey Vandermast's renewed order to destroy the tree, saying "Bacon, that bridles headstrong Belcephon, And rules Asmenoth guider of North: Bindes me from yielding unto Vandermast."[54]

Vandermast accuses Bacon of using "more than art" to win the contest, and he is rewarded for the insult and his limited imagination by being conjured back to Hapsburg, leaving Bacon to receive Henry's praise: "Bacon, thou hast honoured England with thy skill, and made faire Oxford famous by thine art." Bacon's victory over Vandermast, which results in the very symbolic banishment of the loser to his German university to continue his study and reconsider his ideas about the boundaries of art, reflects well on England. It indicates that the nation is sufficiently stable and wealthy to attract, keep, and support unsurpassed masters of the occult arts, and that the king values the pursuit of knowledge as well as the pursuit of land and military honor.

Bacon is not content to allow England to reap all the glory of his victory, proving that he is not completely subjugated to the nationalist ideology of the royal court but instead pursues the study of nature to satisfy his own curiosity and legitimizes and protects his study by making it relevant to the king's goals. He makes his point when, after inviting the king and his entourage for dinner, he presents them with his usual evening meal, replying to the German emperor's complaints with "Content thee Fredericke for I shewd the eates To let thee see how schollers use to feede: How little meate refines our English wits."[55] Having demonstrated that English academics do not enjoy as much royal patronage as Henry would have his fellow leaders believe, Bacon begins to mend the rent he tore in his relationship with the king. In the name of his king, he surpasses his previous demonstration of skill and brings honor to Henry and England by providing a feast full of the bounty of the world.

But Bacon's magic cannot be understood only in terms of his victory over Vandermast or his culinary conjuring. The episode with the brass head determines a transition in Bacon's relationship with natural philosophy from a selfish quest for knowledge to a nationalist and moralized one. In fact, it nearly ends his career by forcing him to acknowledge the limits of a single person's ability to manipulate cunning preternatural forces and results in his abjuring the study of nature forever as amoral and selfish. He will, of course, renege on that promise, but his conquest will result in a new approach to the study of nature. He will continue to push the limits of art, but he will do so with the might of England, rather than simply his own intellect and will, to support him. The transition depends upon a shift in intellectual and social space, in which Bacon leaves the academic community represented by the Oxford scholars debating the appropriate limits of the study of nature for Henry's court, which condones all investigations that can be brought to benefit the state. In this new arena, intellectual skill is a source of national pride and defense, and it is displayed and pursued at the will of the king.

Despite his extensive knowledge of the hidden mysteries of the world and his ability to transcend the previously acknowledged limits of natural philosophy, Bacon remains human. His great experiment, the brass head that will ensure the protection of England and his fame, fails because he is too exhausted to spend another night watching over it. His hapless student, alarmed and overwhelmed by his inherited task, falls asleep and fails to wake Bacon when the head speaks for the first and second time, only calling him the third time, after the head is shattered by a preternatural hand. Bacon blames himself for falling asleep, rages at Miles for failing to wake him, then blames the jealousy of gods and demons for staying Miles's hand or sending him such an idiot student. He cries, "But proud Asmenoth ruler of the North, And Demegorgon maister of the fates, Grudge that a mortall man should worke so much, Hell trembled at my deepe commanding spels, Fiendes frownd to see a man their ouermach."[56] Bacon rages against the limits of knowledge, having finally found them, not in his ability to manipulate preternatural forces, but in the frailty of his body, which by requiring sleep left the door open and unguarded against the retaliation of the very forces he thought he had rendered servile.

He retires to his cell, considering the end of his fame and the cost of his own limitations. Bungay finds him alone and encourages him to remember that his reputation was built on more than the brass head, but Bacon is inconsolable and prophecies that he will commit a terrible act before the end of the day. In fact, he does nothing more than what he had previously done for Edward: He uses his crystal to show two young men (the sons of Margret's prospective farmer-suitors) what is happening in Fresingfield. The boys watch their fathers duel to their mutual death, then turn on each other and perish. Bacon, who had jested with Edward when he threatened to kill Lacy for kissing Margret (a scene he watched through the crystal), that "Twere a long poniard my lord, to reach betweene Oxford and Fresingfield," suddenly realizes the costs of interfering with human affairs, particularly those relating to masculine honor, when their swords can reach each other.[57]

The loss of the young men's lives causes Bacon to reassess the effects of magic on the order of the natural world, an order he believes to have been designed by God. By using art to enlist the help of demons to influence the natural world, he had sealed his fate in hell through the selfish pursuit of knowledge and power. He destroys his crystal and vows to his friend, "Bungay Ile spend the remnant of my life in pure deuotion praying to my God, That he would saue what Bacon vainly lost."[58] His vow lasts only until the wedding of Elinor and Edward, when the king asks him why he is so silent, and he responds with a reminder of his repentance, followed by a veiled portent of the joy that will come to England as a result of the marriage.

The king, neither moved by his friar's vow to abstain from magic nor patient with vague prophecies, commands him to reveal the nation's future. Bacon responds to the king's wishes, beginning his prophecy of war, victory, and a lasting and profitable peace with "I finde by deepe praescience of mine art, Which once I tempred in my secreat cell."[59] The transition from the secret cell to the open air of the court is the key to understanding Bacon's uniquely powerful role as a natural philosophical authority. The new Bacon will practice his art for England's sake under the eyes of the king, where his skill will support the defense and expansion of Albion and can truly be claimed as a national treasure. This transition frees Bacon to pursue the art he had always loved and introduces the audience to an acceptable set of reasons for pursuing natural philosophy. It also reinforces Bacon's reputation as a legitimate natural philosophical authority known for stretching the limits of art and the definition of nature to include invisible forces that can be summoned to do human bidding. That reputation would earn him a position on the title page of *The Mirror of Alchimy*, a book of secrets published in 1597 by Thomas Creede, which capitalized on his newly reinforced place in the academic and popular imagination.[60]

This book adopts Bacon as an established authority and assigns him as the author of two treatises that propose extending the study of natural philosophy to include alchemy and the manipulation of occult forces. The other three treatises are canonical within the discipline of alchemy, beginning with the Emerald Table of Hermes Trismegistus and including Hortulanus's commentary on it as well as Galid's alchemical treatise. The book has merit on several fronts: It presents a series of treatises addressing the structure of authority and the place of secrecy in natural and occult philosophy; it proposes a set of desired characteristics for readers and practitioners of secrets; it contains multiple theories about the way natural processes work; and it acts as a practical guide for average readers interested in understanding and manipulating those processes.[61]

The examination of authority begins on the frontispiece and continues through the final essay in the collection. The frontispiece announces that the text is "Composed by the thrice-famous and learned Fryer, *Roger Bachon,* sometime fellow of Martin Colledge and Afterwards of Brasen-nose Colledge in Oxenforde."[62] This sentence is a labyrinth of scholastic attributions, some of which would have been more recognizable than others. Before I begin dissecting them, though, it is worth noting that none of the descriptors allies Bacon with the brass head, the feat for which he was probably best known. The first, and probably the most likely, is that readers would have been attracted to a Bacon text because of the brass head and required other assurances of his occult expertise rather than a repetition of one they already knew. The second is that the story of the brass head revolves primarily around conjuring, an art of doubtful status and one commonly associated with charlatans.[63]

A final possibility is that the mention of Brasennose college would have prompted readers to remember the myth of the brass head, since it was said to have taken place there, and stood in as an academic association and a reminder of Bacon's most renowned project. The very name *Brasennose* brings to mind a brass head, and it intimately links Bacon to Oxford, which had an extensive reputation for producing great English academics that transcended the ranks of those who could afford to send their sons into its hallowed halls. The final word worth noting in the introduction of the central figure in this text is *friar.* Religious men had been the keepers of the academic keys during the medieval period, so they had a legitimate association with learning, particularly in natural philosophy.

The descriptor *thrice-famous* held a great deal of significance for the alchemical community, since it was a near translation of Hermes' epithet, Trismegistus, the mythical Egyptian father of alchemy whose name was applied to the Emerald Table, an alchemical text adopted by medieval and early modern practitioners as foundational.[64] By the late sixteenth century, the legend of Hermes Trismegistus had grown sufficiently popular to have left the confines of the alchemical community and entered popular culture, particularly in the arena of popular natural philosophy. Bacon's authority as a master of the boundaries of natural philosophy is enhanced by his nominal alliance with Hermes, an alliance underlined by the inclusion of the Emerald Table in *The Mirror of Alchimy* immediately after Bacon's first essay.

The next example of the use of authority to reinforce the information in and utility of this text comes in the introduction to the Emerald Table, in which the myth of Hermes Trismegistus and his text is told in full and his role as the father of alchemy is made evident and more exotic. "The Emerald Table of Hermes, Trismegistus of Alchimy. The wordes of the secrets of *Hermes*, which were written in a Smaragdine Table, and found betweene his hands in an obscure vaute, wherin his body lay buried."[65] Readers unfamiliar with the story of Hermes Trismegistus would learn it and, along with it, the deep roots of alchemy. The humanist movement, the tradition dominating academic efforts in the sixteenth century, urged a return to original texts, and the occult arts in particular insisted that the oldest works were the closest to revealing the truths of the world.

The story of Hermes Trismegistus as it is told here adds an element of romance and secrecy to the established aura of authenticity, since the Emerald Table was supposed to have been recovered from a tomb. The secrets revealed in a long-lost treatise written by an ancient scholar and discovered clutched in his dead hands seem to have the stamp of originality and authority required to render them as authoritative revealed truths. The widespread belief among elite practitioners that he had handed down the secrets of alchemical theory and practice would have also contributed to his sixteenth-century reputation as a legitimate authority. Should readers not have been aware of his relationship to alchemical knowledge before they picked up this text, it is reinforced by his commentator, Hortulanus. In his introduction to his interpretation of the Emerald Table, Hortulanus declares his subject "the Father of Philosophers" and begins his task with a prayer to God to help him correctly interpret the text "that by the knowledge which thou has giuen mee, I may bring my deare friends from error, that when they shal perceiue the truth, they may praise thy holy and glorious name."[66]

God, the ultimate authority, appears here as having already told the secrets of his creation to Hermes Trismegistus. Hortulanus asserts his own authority and justifies his commentary by claiming that the Emerald Table has been widely read but poorly understood by a broad cadre of medieval intellectuals. He also establishes himself as a recipient of divine inspiration, claiming that it allowed him to penetrate the inaccessible original text: "yet have I truly expounded the whole operation and practice of the worke: for the obscuritie of the Philosophers in their speeches, dooth nothing preuaile, where the doctrine of the holy spirit worketh."[67]

Readers who arrived at the beginning of the next text would have been reminded again of the weight of the authority behind the theories and arguments made in *The Mirror*. Galid supports his treatise by claiming divine instruction in the creation of the philosopher's stone earned through years of study and diligence. "Know brother, that this is our mastery and honourable office of the secret Stone, is a secret of the secrets of God, which hee hath concealed from his people, neither would he reveale it to any save to those, who like sonnes have faithfully deserved it, knowing both his goodnesse and his greatnesse."[68] This sentence accomplishes two things for the author: It establishes his credentials as a serious student of God and nature, rather than a charlatan, and it defines his role as a translator of God's secrets for those who lack his skills and experience.

Many natural philosophers and theologians believed that the ancient Hebrew fathers like Abraham and Moses and a few of the ancient Greeks, particularly Plato and Aristotle, received the truths of the world from God as a reward for their exemplary lives and morality. This sentiment is repeated in the final treatise in this collection, supposedly written by Bacon, in which he reviews the means by which ancient scholars have kept their secrets: "and therefore there is a great concealing with them, but especially with the Jews: for Aristotle sayth in the above named booke [Book of Secrets], that God gave them all maner of wisedome, before there were any Philosophers, and all nations borrowed the principles of philosophy of them."[69] Those truths were considered lost to contemporary Christians, who lacked the sanctity to earn God's trust, and so ancient texts were their best hope for gaining access to lost secrets.[70]

The final essay also reaffirms Bacon's ties to Oxford. Its subject matter, a theoretical treatise on the difference between the insubstantial tricks of magic and the remarkable force exerted by art and nature, establishes *The Mirror* as part of an ongoing debate in natural philosophy that occupied sixteenth-century thinkers inside and outside the academy. This treatise enters that debate about the ideal nature of authority by offering established names and texts to support the primary arguments, repeatedly citing Aristotle and Artephius, Pliny, and Ptolemy, but challenging the place of inherited authority and forwarding the importance of personal experience in a scholarly endeavor such as natural philosophy, in which the boundaries of knowledge are always changing. "For we knowe that *Aristotle* sayth in the *Predicaments* that the quadrature of a Circle may bee knowne, although it been not yet knowne. Whereby hee confesseth, that both himselfe, and all men till his time were ignorant of it. But now a dayes wee see that the truth is knowne, so that Aristotle might well be ignoraunt of the greatest of Natures Secrets."[71]

The tension between depending upon the claims made by ancient authorities and highlighting examples of knowledge developments that render the same authorities obsolete captures the predicament facing both academic and popular natural philosophy. Book sales depended upon a book's ability to contribute something new to the body of common knowledge, but its ability to make those claims rested on its portrayal of authoritative texts. All three theoretical treatises, the two attributed to Bacon and the one attributed to Galid, share approaches to this problem—they all define, to some extent, problems with existing texts, they police the boundaries between academic and popular knowledge, and they provide practical, step-by-step advice that would guide readers through the creation of the pinnacle of alchemical practice, the philosopher's stone.

Readers might have been attracted and reassured by the repetitions of authority and authenticity, but they would have been most interested in the information revealed in the five treatises. Nothing demonstrates this better than the careful notes in the margins of the copy of this book held by the British Library, which outline the contents of the first and last treatises. Occult texts were known for their coded and generally opaque writing style, and the treatises in *The Mirror of Alchimy* possess a splendid combination of clarity and confusion.[72]

The three theoretical pieces and the commentary agree that secrecy or obscurity have long been hallmarks of ancient authoritative texts. In the preface to the first text, "Bacon" writes, "In times past the Philosophers spake afters divers and sundrie manners throughout their writings, sith as it were in a riddle and cloudie voice, they have left unto us a certaine most excellent and noble science, but altogether obscure," and in the last piece, "Now the cause of this concealment among all wise men is the contempt and neglect of the secretes of wisedome by the vulgar sort, that knoweth not how to use those things which are most excellent."[73] These two excerpts contribute to the idea of ancient knowledge having been deliberately obscured to protect it and its creators/discoverers from the attentions of the masses.

In "The Admirable Force of Art and Nature," from which the second excerpt is derived, the author goes on to catalog the variety of methods writers have used to disguise critical pieces of natural knowledge, rendering it inaccessible to all but the most devoted and educated students. He notes the prevalence of characters, verse, figurative speech, codes, mixtures of letters from different languages, the use of foreign languages, symbols that stand in for words, and "art notory."[74] His relationship to that secrecy is made clear immediately after the catalog, when he claims, "I deemed it necessary to touch these tricks of obscurity, because happily my self may be constrained through the greatnesse of the secrets, which I shal handle, to use some of them, that so at the least I might helpe thee to understand."[75] By making evident the methods of textual concealment, the author provides for readers unacquainted with these techniques a framework for interpreting the information he provides while, at the same time, preserving its identity as "secret" by keeping it hidden. He even highlights the pieces that require extra attention from the reader with phrases like, "Now I will hide another secret from thee," drawing the eye and the mind to the information that follows this introduction.[76]

The remaining texts in this collection engage as much in secret keeping as in secret telling, and they make use of the easiest strategy of obfuscation outlined above, the use of highly figurative language. "The Myrrour of Alchimy" is full of thick sentences like "And for as much as nature doth always work simply, the perfection which is in them [gold and silver] is simple, inseparable, & incommiscible, neither may they by art be put in the stone, for ferment to shorten the worke, and so brought to their former state, because the most volatile doth ouercome the most fixt."[77] The key to unlocking this text is not presented in a neat list but is instead embedded in the work itself. At the end of his theory of coagulation, for example, the author directs readers back to the beginning: "Be therefore wise: for if thou shalt be subtile and wittie in my Chapters (wherein by manifest prove I have laid open the matter of the stone easie to be knowne) thou shalt taste of that delightfull thing, wherein the whole intention of the Philosophers is placed."[78]

The recommendation to reread would have fit in particularly well with the kind of reading historians believe to have dominated the early modern approach to texts. "Intensive reading" involved multiple rereadings of books and careful attention to language, so that readers could grasp the full range of the information presented to them. It also encompasses the multiple venues of early modern reading, including the role of reading texts aloud as a means of instruction, entertainment, and enlightenment.[79] In this case, the instruction to reread before continuing with the text would have been crucial to understanding the theory behind the practical advice provided later on, but it is not clear that the practical pieces would have been impenetrable without it. This suggests the importance of another delineation made in this collection of texts, between people truly interested in grasping the breadth and depth of their subject and those simply interested in mastering the technique for personal gain. The latter strategy comes in for some harsh criticism and the repeated promise of failure—it appears that one of the factors keeping secret knowledge secret in this book is readers' willingness to engage with the difficult, encoded, or opaque language in order to master the theory necessary to correctly interpret the more plainly presented recipes for the philosopher's stone.

While all these treatises acknowledge the prevalence of secrecy in ancient texts and generally praise it as necessary for shielding natural mysteries from the prying eyes of the unworthy, they also share the task of telling those secrets. The commentary by Hortulanus, for example, exists simply to decode the language in the Emerald Table and, in the words of its introduction, to allow the author the chance "that by the knowledge which thou [God] hast given mee, I may bring my deare friends from error."[80] The author of "The Booke of the Secrets of Alchimie" promises to conceal nothing in his clarification of his predecessors' works, claiming that his pupil Musa's endless struggles and despair after his attempt to master the obscure mysteries of ancient texts and reconcile their arguments drove him to write this work on his deathbed.[81] He also promises to reveal them in language suited to a particular audience composed of "him that is studious of this Art or masterie, in a language befitting his sence and understanding."[82]

This depiction of the desirable reader as a man interested in alchemy but lacking the desire and the skill to drive himself to despair perusing an array of original texts in various languages written with the intent to confuse, prevails through all the essays in this little book. Diligence and patience, much more than education, describe these readers, who were expected to intensively interact with each treatise, undergoing a process of intellectual enlightenment and purification that mirrored the stages through which the stone passed on its way to perfection.[83] Diligence is the hallmark of the desired reader, who, upon following Galid's instructions for making the philosopher's stone, will spend at least 110 days perfecting his creation.[84] This commitment to the text and to the work required to produce the elusive philosopher's stone was an intrinsic textual device that separated true "readers of secrets" from dilettantes and contributed to the belief in the secrecy of information presented in relatively accessible, inexpensive form to a broad audience in a vernacular language.

Readers were also expected to be canny and aware of the tricks used to legitimize charlatanry and the various properties of magical and artful or natural acts. In a moment of high irony in the final treatise, fully entitled "An excellent discourse of the admirable force and efficacie of Art and Nature," pseudo-Bacon points out that magicians and conjurers are known for ascribing their own new and unfounded ideas to great authorities in new books. "And to the ende men might be the more thoroughly allured, they give glorious titles to their workes, and foolishly ascribe them to such and such Authors, as though they spake nothing of themselves, and write base matters in a lofty style, and with the cloke of a text do hide their own forgeries."[85] He warns that wise readers will parse philosophers' texts, separating their realizations of natural phenomena or legitimate arts like astrology and alchemy from the magical incantations, charms, and rituals of frauds.[86]

Pseudo-Bacon also separates his audience into "learned" and "common," dividing them according to their relationship to basic principles of natural philosophy. "I speake of the Common sort, in that Sence, as it is here distinguished against the learned. For in the common conceytes of the minde, they agree with the learned, but in the proper principles and conclusions of Arts and Sciences they disagree, toying themselves about mere appearances, and sophistications, and quirks, and quiddities, and such like trash, whereof wise men make no account."[87] This division, however, does not exclude readers willing to engage with his text, in which he introduces the fundamental ideas behind natural philosophy and provides examples of the relative potential of art and nature as compared to the external rewards of magic. In fact, it produces a definition of wisdom that can be fulfilled solely by careful reading of and agreement with the treatise itself.

The fourth aspect that the treatises in this text share is a commitment to providing models for the ways in which natural processes occur. The presentation of these models supports the alchemical theory that surrounds them, making a comparison between the reactions that happen in the alchemical furnace and the processes of rock formation in mountains. In "The Myrrour of Alchimy," the parallels between the natural world and the alchemical furnace are made explicit. The model is introduced in the context of clarifying the nature and use of the flame in doing alchemical work and points to the natural production of the two ingredients required to make the philosopher's stone, argent-vive [mercury] and sulfur. "Do we not see that in the Mynes through the continuall heat that is in the mountains thereof, the grosnesse of water is so decocted and thickened, that in continuance of time it becommeth Argent-vive? And that of the fatnesse of the earth through the same heat and decoction, Sulphur is engendred?"[88] The author argues that mimicking the natural formation of these two minerals is the only way to produce them artificially, and he uses this to forward his own alchemical process over those proposed by other authors who suggest means other than continual concoction. "Woe to you that will overcome nature, and make metals more then perfect by a newe regiment, or worke sprung from your owne senselesse braines."[89] The natural world also serves this author as a model for ideal alchemical apparatus. He asks, "And if we purpose to imitate nature in concocting, wherefore do we reject her vessel?"[90] He follows this question with a series of recommendations for alchemical tools based on the characteristics of the natural sites in which the processes they are mimicking take place.

The comparison serves as a guide to readers who, searching for the correct equipment to begin their alchemical careers, can refer to the known quantities of mountains, the earth found at their edges, and constant fire. There is also a more theoretical aspect to the parallel between nature and alembic. The natural models summoned at the behest of alchemical theory are Aristotelian, composed of the four elements of water, air, earth, and fire, and characterized by the four qualities of hot, cold, moist, and dry. "The Booke of the Secrets of Alchimie" takes this familiar model and alters it to properties associated with the elements and their qualities, namely the ability to dissolve, congeal, and render something white or red.[91] By enlisting a familiar model of the generative forces prevailing in the natural world, these authors make alchemical transformation, an art, into a natural process driven by known (or at least nameable) quantities and consummated with established outcomes. The alliance between natural processes and alchemy not only legitimizes and makes godly the sometimes questionable work of the alchemist, it also produces a predictable, reproducible model of the natural world.[92]

The outstanding thing about each of the three alchemical treatises in this book is their devotion to practicality. While each expends some effort on the explication of a particular alchemical theory or dissection of the relative merits of art and nature, each concludes with a clear, relatively easily followed recipe for making the philosopher's stone. "The Myrrour of Alchimy" and "The Booke of the Secrets of Alchimie" begin their practical advice with instructions for obtaining the proper equipment. These instructions are anything but opaque; the first treatise defines their characteristics in comparison to their natural analogs, while the second calls them by their proper names, as though issuing a shopping list to the would-be alchemist. "And every one of these must have a Furnace fit for it, and let either of them have a similitude and figure agreeable to the work. . . . As for the instruments, they are two in number. One is a *Cucurbit,* with his *Alembick:* the other is *Aludel,* that is well made."[93]

In conjunction, these treatises produce a workable image of the required equipment and its proper names, so that a new alchemist purchasing his first tools could obtain the right items. They go on to outline the steps that should be followed to make the philosopher's stone, beginning with the repeated dissolving of the two ingredients in "water" and congealing over fire until their natures are joined together and the soul is trapped in the mineral body. The next step involves decoction, contrition, and washing of the stone, "so that he who will bestow and paines herein, must cleanse it very well, and wash the blacknesse from it, and darknes that appeareth in his operation, and subtiliate the bodies as much as hee can."[94] A careful reader would note that the next step in the text is actually required to perform the first step, since it addresses the need to do the initial dissolution and congealing in mineral water, to prevent the ingredients from being scorched by the flame. Once the stone has been thoroughly washed, it should be divided in the cucurbit according to the four elements. The solid that falls to the bottom of the glass should be heated over a gentle flame until the blackness disappears and it turns white. It should then be transferred back to the cucurbit, covered with an alembic, and buried in hot dung to allow distillation to occur.

Once all the moisture is gone, the stone should be removed, ground, and put in an appropriate vessel, then buried in very hot horse dung, which is constantly replenished with new supplies of hot dung, and left for forty days, by which time it will become a thick white water. That substance should be weighed, then half its weight substituted with the water taken in the original distillation, then set again in hot dung for ten days. That step should be repeated five more times, for a period of ten days each, until the stone is perfected. The proof of success lies in the application of the completed stone to two hundred fifty drams of lead or steel, which it should convert it to silver or gold.[95]

Conclusion

This chapter has argued that the liberal attribution of natural knowledge to classical and medieval authorities with established credentials as masters of the natural world was a crucial aspect in determining the legitimacy and popularity of books of secrets. Authorities could be entirely fictional, like the wandering Jew Erra Pater, touched up and attached to popular sixteenth-century concerns such as nationalism and defining the limits of art and nature like Roger Bacon, or simply expanded to fill the position required of them, as demonstrated by Aristotle's dominance in the area of popular gynecological and obstetrical texts. Regardless of their derivation, they served to support the claim made by all books of secrets, that the knowledge they were presenting came from the pens of men who had extensive reputations for mastering the natural world. The diffusion of figures like Aristotle, Pliny, Magnus, and Bacon into the market for books of secrets indicates that the border between academic and popular culture was open, and that figures with widespread academic recognition were also good bets as popular authorities. Furthermore, the presence of books of secrets attributed to figures such as Albertus and Bacon on the shelves of educated men like John Dee reiterates the extensive shared space between learned and popular culture and suggests that early modern people, regardless of their level of formal education, shared an interest in using art to expand their control over nature.

The specific case of Roger Bacon also points to the extensive intersections among different aspects of popular culture. Roger Bacon first emerged as a figure of interest at court and in the libraries of educated men who collected works attributed to him, as well as some written by him, in hopes of manipulating their fortunes through alchemy. Robert Greene, a playwright whose university education differentiated him from many of his peers, appropriated Roger Bacon as a legendary medieval magus and wrote a play that made him into a moral object lesson about the appropriate reasons and means to pursue dominance over the natural world. In the process, however, he established Bacon's academic credentials and introduced him as a reliable authority, if not a man in complete control of human weaknesses such as pride and selfishness, in the arena of natural philosophy. That image was supported by the printer of *The Mirror of Alchimy*, who purchased the rights to print the book (it was translated from the French) in hopes of capitalizing on Bacon's recent reintroduction to popular culture. Between the play and the book of secrets, the image of Bacon became a staple of English popular culture in the seventeenth century, though he was separated from the serious academic slant and the particular natural philosophical agenda of *The Mirror of Alchimy* within several decades by changing interests and obligations in natural philosophy. The arena of print about and print attributed to Bacon lasted well into the seventeenth century, spawning titles like *The Most Famous History of the Learned Fryer Bacon* in 1624 and *Frier Bacon His Discovery of the Miracles of Art, Nature, and Magick* in 1659.

The next chapter will move on from the attribution of popular culture and the creation of authorities of secrets to address the role that textual structure played in making books of secrets attractive to readers and introducing them to specific models of the natural world. This chapter will form a bridge of sorts between printers and readers, contending that printer expectations of reader interests largely determined the ways in which books of secrets were packaged, and that the structures that were applied to them made them usable in two very different ways that propose a new model for thinking about early modern reading practice. It will also argue that books of secrets contained very specific instructions for the types of readers who would be most capable of benefiting from the information they contained. Those instructions specified diligent, careful readers, just like those described in *The Mirror of Alchimy*. They also introduced a new kind of authority within books of secrets—the readers', and they established readers' experiences as parallel to and capable of informing those of lauded authorities like Aristotle and Bacon. This construction of good readers accomplished two things. First, they were made a part of the process of thinking about and categorizing the natural world, and second they were made complicit in the larger project of classifying the information presented in books of secret as secret. Under these conditions, the definition of secret changed to mean that only those who could correctly use the knowledge actually knew it. And because they were constrained by their discipline and sense of responsibility not to share what they knew with the uninitiated, they were made colleagues in secrecy with the masters of the natural world to whom these works were attributed. This collapsing of identities would be critical to the formation of an audience for books of secrets that was unified by its subscription to the belief that they had access to protected knowledge that permitted them to manipulate nature to their own benefit.

CHAPTER THREE

Structuring Secrets for Sale

"First therefore I wil declare of certaine herbes: Secondly of certaine Stones, and thirdly of certayne Beastes, and the virtues of them."[1]

"Manie are the wonders and marvels of the world, and almost incredible, were it not that experience teacheth the contrarie:"[2]

The last chapter established the creation of authority as a fundamental part of books of secrets, since their appeal depended on readers believing that they contained the knowledge of great masters of the natural world whose wisdom was suddenly and for the first time available at a price they could afford. It also introduced the responsibility given to readers of secrets, using the model reader created in *The Mirror of Alchimy* to establish the importance of characteristics such as diligence and patience. This chapter will elaborate on the role of the reader of secrets in two ways. First, it will provide a detailed structural analysis of two general books of secrets, *The Book of Secrets of Albertus Magnus* and *Cornucopiae, Or divers secrets*. These books are ideally suited to the task because they presented complete models of the natural world, rather than focusing on a specific aspect of nature or a particular art. This means that their internal structure influenced the kinds of natural worlds that readers imagined themselves trying to change. Furthermore, the kinds of worlds they depicted were not always in perfect congruence with the text, which raises interesting questions about the ways that readers interacted with their books of secrets. Second, it will expand the definition of a good reader of secrets by establishing that both the structure and the text of these two books validated and legitimized readers' experience. Thus, they both empowered readers and strengthened their position in relation to nature and added a new type of authority to those explored in the last chapter.

The books of secrets that have attracted the most scholarly attention are compendia of natural knowledge that reflect the relationships among natural objects and offer the reader methods for manipulating nature or producing illusions. Two of these books in particular are notable for their accessibility to a broad audience and their different approaches to organizing very similar information about the natural world. *The Secrets of Albertus Magnus* and *The Cornucopiae, Or divers secrets* are remarkable because they contain elements of natural knowledge that had previously been printed only in Latin, a language that would have been inaccessible to the majority of readers, and because they would have been relatively affordable. Their page count and typeface locate their price between three and six pence, the price depending upon the price of paper and the viability of the print market.[3] These books provide insight into the techniques used to contain and present natural knowledge to this segment of the print market, since they represent markedly different strategies for doing so. One of these strategies depends upon headings and tables of contents to categorize the information contained in the text. The addition of a structuring framework does not, however, invalidate the potential for other structural models embedded within the text. *The Secrets of Albertus Magnus* contains both an imposed and inherent structure, and the two offer readers very different ways to interact with the text and with nature itself. *Cornucopiae, Or divers secrets* does not have an external structure at all but is instead composed of repeating sections of interconnected natural objects whose relationships determine the ways that change can occur in nature. Rather than experiencing nature divided into artificial categories, a strategy that did not succeed even with the best efforts, readers of *Cornucopiae* would have encountered the natural world as it was constructed through the relationships between items, as well as gaining entrance to a natural world whose predictability, once apprehended, made it subject to human manipulation and desires.

The Secrets of Albertus Magnus saw four English printings in the sixteenth century and features individual books on the virtues of herbs, stones, and beasts, and one devoted to miscellaneous wonders. Each book is divided from the others with clear headings in large type, so that readers would have found it difficult to miss the transition from one natural category to another.[4] The 1599 edition produced by William Jaggard includes an additional book on the movements of planets and their effects on children, and it also features the most dramatic divisions among the sections of the text.

Despite the effort devoted to assigning the knowledge presented in *The Secrets of Albertus Magnus* to specific categories and developing those categories into a framework for an orderly natural world, the knowledge in this text resists organization. The structuring categories are undermined by the mutability of the book's content, with entries sometimes appearing more than once in different sections and some examples occurring well outside their "logical" location. The tension between the extrinsic structure, as found in the headings, and the intrinsic structure, which resists and undermines the headings applied to it, allows for a closer examination of the book's underlying organizing principles. Those principles depend more upon links between natural objects, particularly relationships of sympathy and antipathy, than a specific object's placement in a broader category like stones, herbs, or beasts.

The *Cornucopiae, Or divers secrets* demonstrates this structuring technique and presents a natural world that is the sum of the relationships between its many parts. This book, which was printed in 1595 and 1596, is an excellent example of a text-specific and text-dependent system for thinking about the natural world. No categories, or even recognizable headings, organize it; in fact it appears to have no structuring principles at all. But a careful reading uncovers the structure, in which the content of one entry determines the content of the next, and systems of sympathy tie the world together.[5] These systems of sympathetic relationships produce a web of linked elements that lend insight into the way early modern readers could have thought about and experienced the natural world, both in print and in reality.

This chapter will compare the systems and categories used in *The Secrets of Albertus Magnus* and *Cornucopiae, Or divers secrets,* with an eye to uncovering their organizing structures and the worlds those structures offered to readers. It will attempt to uncover the diverse ways in which both systems made it possible for readers to make sense of the texts and the world around them. Both frameworks produced texts that allowed readers to construct two different models of the world: local or personalized natural worlds whose existence depended on interest in or experience with the listed items, as well as more general natural worlds that derived their shape from accepted and broadly recognized categories. The *Cornucopiae, Or divers secrets* offers a place from which to start this discussion, since readers would have had only its internal organizational system by which to navigate the natural world it produces.

The fundamental principles underlying the organization of this text include the assumption that every natural object has a set of inherent characteristics that determine its relationships with other objects, and that those relationships can be characterized as sympathetic or antipathetic. They are construed as the active force in nature, and the point at which nature is most vulnerable to human manipulation. The structure of the *Cornucopiae* depends upon the explication of relationships between objects and is arranged in a list of linked entries, which establish the ties between apparently dissimilar objects through their connections to other objects that the first two have in common. These chains of entries, which are tethered together by characteristics and dependent upon each other for their meaning, indicate the continued importance of sympathy in popular depictions of the relationships governing the natural world.

The complex natural models in these books reflect the dedication required by readers to master them. Both *Albertus Magnus* and the *Cornucopiae* devote time to building up images of desired readers and imbuing their experience with the power to support or, in the case of the former text, challenge the secrets they will encounter within the text. The empowerment of readers' experience acknowledges the importance of their subjectivity in determining the lives they bring to bear on books of secrets, a factor that is also reflected in the implicit structure of both texts. This chapter will interrogate the connections between textual structure and the position of readers' authority and experience to produce a richer understanding of the ways in which books of secrets presented personalized, navigable, and predictable images of the natural world.

This chapter will question the ways in which authority, secrecy, and experience are represented in *Albertus Magnus* and *Cornucopiae,* with particular attention paid to the ways in which the textual structure supported readers' authority and the books' claims to secrecy. The importance of authority and its relationship to experience cannot be underestimated in books of secrets, because their success depended upon convincing readers of three things: that the information they contain is truly secret or novel because it comes from recognizable authorities whose knowledge has been almost entirely unavailable to most readers; that the information revealed in the books remains secret even after it has been read; and that only special readers with specific personal characteristics like patience, cleverness, and diligence would be able to access and make use of that knowledge.

Theoretical Framework

A thorough structural analysis of these texts depends upon a combination of historiographic and literary critical approaches. Cultural history, in the form of Roger Chartier's idea of appropriation, is fundamental to seeing these texts as spaces of interaction between inherited knowledge and popular conceptions of the natural world.[6] He asserts that early modern readers apprehended texts on the level at which they were or seemed to be most relevant to their lives and concerns. In other words, they made sense and took possession of the knowledge they read according to the questions and matters of importance in their lives, not necessarily according to their social class, their level of education, or how that knowledge was presented or categorized. He stresses the importance of this process for understanding popular reading practices and contends that readers from different social classes could approach the same text and make entirely different sense of it.

Chartier writes, "Above all, the 'popular' can indicate a kind of relation, a way of using cultural products or norms that are shared, more or less, by society at large, but understood, defined, and used in styles that vary."[7] Chartier's argument enriches our understanding of the position occupied by inexpensive books of secrets in the early modern print marketplace in that he allows for a wide variety of readers purchasing and engaging with the same text. This is relevant to a discussion of books of secrets, since they were read in common by a broad cross section of the reading public and frequently appeared in multiple editions, some identical and some with slight variations.

I would like, however, to challenge Chartier's notion that every reader made sense of the text only according to his or her daily life and experiences. I contend that the structure of these books determined a relatively limited position that readers could occupy in relation to the natural world. Readers, once they had chosen to purchase an inexpensive compendium of natural knowledge, would have shared a perspective on the world that transcended differences in social class and educational background. They would have been looking out the same textual window, and the world they saw through it would have depended more upon the structure and content of the text than their individual experiences with or their desires for natural knowledge. This is not to say that personal experience has no place in the reading of books of secrets—in fact it certainly would have contributed to readers' decisions to buy the book in the first place and probably shaped their experiences with it—but the books' structures would have influenced readers' conceptions of the way the natural world functioned, their position within it, and their ability to manipulate it according to their needs. Readers of these books would have shared a perspective that emphasized the centrality of their needs in respect to nature and their ability to control the natural world through the understanding of relationships between objects and planned manipulations of those relationships. This position would likely have been at odds with the one experienced by many early modern readers, who must have found themselves to be nature's victims, whether through disease, bad weather, or crop failure, more frequently than they were its masters.

The piece that Chartier does not address, and the one that supports the above argument, comes from the comparative literature scholar Roland Barthes. He has proposed an alternative to the narrative progression model that has long dominated the way scholars thought about the way people read books.[8] In this model, narrative progression depends upon stable actors identified by consistent names and characteristics that use consistent and accessible language for the purpose of naming things, and linear movement through the text that is nearly impossible to resist, subvert, or avoid. *The Secrets of Albertus Magnus* might have lent itself to this kind of reading, as long as each of the four books it contains was addressed as independent from the others. The work deviates from the idea of narrative progression as a complete volume, however, by breaking up the text among the three books of virtues and the book of wonders. The structure of each book of virtues also resists being read only according to narrative progression, since each one begins with a table of contents that would have allowed readers to enter the text at any point. Furthermore, there is no real narrative in *Albertus Magnus,* at least not as narrative is generally defined. Instead, it constructs of an image of the natural world as predictable and subject to manipulation that repeats throughout the books of virtues. Because of the repetition, this idea of the natural world can be absorbed by beginning to read at nearly any point in the first three books. The book of wonders, however, points to the potential for narrative progression, since the image of a predictable and manageable world culminates in this explication of the interesting things that already exist in nature. The elements in this book are not wondrous because they violate the laws of nature or operate invisibly, as William Eamon contends in his definition of secrets or wonders, but because they demonstrate the potential inherent in the natural world to readers who can appreciate and learn to manipulate it.[9]

The structure of *Albertus Magnus* might lend itself to intensive reading, which has dominated historians' ideas about medieval and early modern reading patterns. The repetitive nature of the books of virtues would, in fact, have encouraged readers to master details about plants, animals, and stones in small chunks while also repeatedly appropriating the idea of a natural world that was subject to specific kinds of human influence. Finally, readers would have absorbed the message that their needs and desires had power in and often over the natural world they were encountering, and the book would have then invited a more utilitarian approach to satisfying specific needs and desires. A reader interested in using marigold as a medical remedy would have been able to satisfy her interest without having to read the whole book, though she would have missed out on any information on marigold contained in the books of herbs and beasts if she had done so, since the tables of contents contain only entries in the category to which they are devoted. *Cornucopiae,* on the other hand, would not have been particularly useful as a reference book unless readers had already spent a great deal of time with it and separated its webs of connections into smaller subsections that they could annotate themselves. While there is evidence of reader interaction with this book, including notations next to information suggesting that the reader found the contents either interesting or true, I have not yet found evidence of that kind of textual demarcation.[10] I believe that *Cornucopiae* was designed to be read as a guide to the whole world, with each of its unmarked subsections producing smaller iterations of nature that could be put together to form a coherent image of the whole.

Cornucopiae resists any reading that depends upon narrative progression or requires it to answer direct questions about the virtues of a specific natural object. Like many early modern works, it has an open structure that has the potential to repeat, be reversed, and circle in on itself from any point. Roland Barthes argues that books like this one move beyond the simple act of naming things to the more reader-determined and interactive strategy of playing with signifiers, linking them in complicated chains of related words and characteristics that challenge the notion of a singular subjectivity and propose definition through difference or through multilayered determinations of identity.[11] *Cornucopiae* is an excellent example of this kind of book, since it consists of an unmarked expanse of text in which each element introduces the next and the context surrounding each element changes according to where you begin to read. Some elements appear multiple times and occupy multiple positions, but none of those positions depends upon picking up the book at a certain point or reading it straight through. Furthermore, none of them is defined so tightly that it cannot be located in another place, in relationship to a different entry or chain of entries. The book is structured to be entered and left at any point, and since each chain of entries reiterates that the nature can best be understood according to the relationships between natural objects, it produces identical frameworks for nature approximately every three pages.

Borrowing analytic tools designed to deconstruct literature in order to understand the place of structure in books of secrets may seem to be an exercise in interdisciplinary futility, but it is a productive strategy because it uncovers some of the expectations that influenced a diverse array of literary products. It also undermines the primacy of the modern labels "popular" and "elite" that have been laden with a multitude of meanings and attached to early modern books. Edmund Spenser's opaque epic *The Faerie Queen* is an excellent example of the ways in which this strategy can be useful. Despite its reputation and elegance, this epic poem is also suited to a reading like the one required to get through *Cornucopiae*, a book for which we have few scholarly expectations. Once you are willing to consider *Cornucopiae* and *The Faerie Queen* as similar—both were after all produced to sell copies and share a desire to communicate specific ideas about the relationship between people and the world they live in—the early modern print market gets both larger and much more flexible. And that flexibility better reflects the purchasing and reading habits of early modern readers than do the highly modern divisions that scholars have applied to the print market to make it and early modern readers suit our expectations.

Printer Strategies

One of the most compelling aspects of working with translated and compiled texts is the opportunity to think seriously about the structuring efforts of translators and compilers, who were often the same person, and printers. The decisions compilers and translators made include what knowledge to include, what to leave out, what kind of vocabulary to employ, and how to present it as part of a complete package that printers would want to purchase and license. A very different kind of study would have to be done in order to understand those efforts, and the results would likely reflect highly individual approaches to this task, since translators and compilers were not as organized a group as were printers.[12] I will turn my attention instead to the structuring decisions made by printers, who faced many of the same questions outlined above. In order to thrive, they had to structure the information in the books they purchased so that it met their conceptions of the needs of a markedly heterogeneous audience—the reading public. The decisions they made about pleasant and functional organizational models reflect their ideas about the type of books that would resonate with and have relevance to a multitude of readers. They also permit insight into the ways in which readers could have experienced these books, since the structures of the texts enable some approaches while making others difficult to maintain. Finally, they underline the fact that books of secrets were part of a much broader print market—their frontispieces and organizational structures are not unique in the world of inexpensive print, and thus they reflect strategies that printers frequently employed to attract readers and sell books.

The Secrets of Albertus Magnus is an excellent place to begin thinking about printer strategies, since its multiple editions are remarkably consistent in the ways in which they made new knowledge accessible to readers. The task of constructing a reliable source of secrets begins on the title page, where all four sixteenth-century editions use the same strategy for drawing readers' eyes from the large print at the top of the page through the first and into the second subtitles. They present the title, *The Book of Secrets of Albertus Magnus*, in the shape of two reversed pyramids with the largest type appearing on the first line and a space of several lines between the end of the first pyramid and the beginning of the second.

The second pyramid is also the beginning of the second subtitle, a distinction made visible through spacing instead of a shift in the size of the type. Readers' eyes are not simply abandoned at the end of the type; both the 1560 and 1565 editions feature decorative borders around their titles, with the most elaborate decoration reserved for the bottom of the page in the shape of bold, horizontal designs that move the eye across the page and to the vertical strips of design that frame the title, drawing attention back to the top of the page and its large print. (See Figures 1 and 2).

The most dramatic departure from this design occurs in 1599 in the edition printed by William Jaggard (Figure 3). In this edition, the title changed from "the book of secrets of Albertus Magnus" to "the secrets of Albertus Magnus," a substitution made more apparent by the use of visual strategies similar to the Seres edition for the organization of the frontispiece. Once again, it is organized into separate sections loosely arranged into reverse pyramids that draw the eye from the largest typeface in the first two lines through the array of subtitles to rest on a well-placed ornate woodcut, then the printer's name and the date of printing. This frontispiece is absolutely deliberate in its manipulation of readers' eyes. The largest print is reserved for the catchiest phrase, "THE SECRETS/ of Albertus Magnus," which appears at the very top of the page where it is most likely to attach a browser's attention.

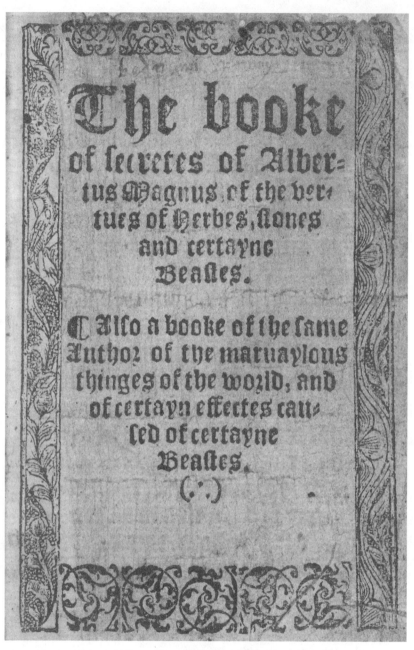

Figure 1. Frontispiece of the 1565 edition of *The booke of secretes of Albertus Magnus of the vertues of Herbes, stones, and certayne beastes*, published by William Copeland. Courtesy of the Huntington Library Art Collection and Botanical Gardens in San Marino, California, where it is part of the Rare Books Collection under call number 17106. Further reproduction is prohibited without permission.

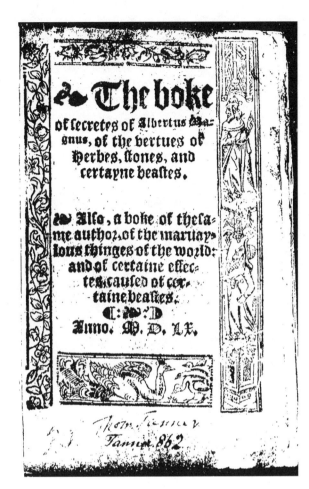

Figure 2. Frontispiece of the 1560 edition of *The boke of secretes of Albertus Magnus, of the vertues of Herbes, stones, and certayne beastes*, published by John King. Courtesy of the Bodleian Library, Oxford University, Antiq. F.E91 and published with permission of ProQuest Information and Learning Company. Further reproduction is prohibited without permission.

The privilege previously reserved for the material object of text has been reassigned to the more ephemeral "secrets" promised here. And these are not just any secrets, but those of Albertus Magnus, whose name carries a great deal more weight here than it did before—it appears whole for only the second time (the first was 1570) and, although not entirely capitalized, it is printed in type nearly as large as that of "The Secrets" themselves. Hooked by this tactic, readers' eyes would drop downward over a skipped line and arrive at the first subtitle: "Of the Vertues of Hearbes,/Stones, and certaine Beasts" covers two lines, with the first one appearing in larger type than the second. Emphasis is again given to the attention-catching phrase, "Of the virtues," which reflects the nature of the text's content and invites further reading by being recogniz-

Structuring Secrets for Sale / 69

Figure 3. Frontispiece of the 1599 edition of *The Secrets of Albertus Magnus Of the Vertues of Hearbes, Stones, and certaine Beasts,* published by William Jaggard. Courtesy of the Wellcome Institute for the History of Medicine, London and published with permission of ProQuest Information and Learning Company. Further reproduction is prohibited without permission.

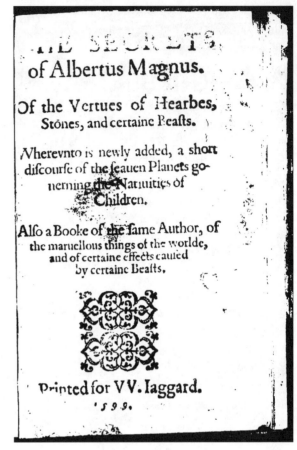

ably useful, given the appeal of herbs, stones, and beasts as buffers between humans and disease or weakness.

The second significant change in the 1599 edition of the text came with the addition of a new section concentrating on astrology and the planets' effects on children. While the first subtitle could attract readers who had not yet purchased *The Secrets of Albertus Magnus,* the second subtitle offered something to those who already owned another edition or who believed they knew enough about the secrets of nature. This subtitle has central billing on the page and, despite its smaller type, is nearly impossible to miss, since it takes up four lines and makes dramatic use of the reversed pyramid to bring readers from its first to its last lines. It follows another skipped line and, in even smaller print, offers: "Whereunto is newly added, a short / discourse of the seauen Planets go-/ verning the Natiuities of / Children." "Children" stands on its own line, perfectly centered and capitalized, as if to emphasize why this new section is relevant. The knowledge it presents concerns children, an aspect of nearly every reader's life, and it offers parents a system for predicting their infants' personalities and habits, or for retrospective analysis of the successes or failures of older offspring.

The final subtitle of *The Secrets of Albertus Magnus* is dramatic only in its organization, which is once again in the shape of a reverse pyramid. The first line of the final subtitle shares the same size type as the one just described, but its concluding lines are even smaller. "Also a Booke of the same Author, of / the maruelous things of the worlde, / and of certaine effects caused / by certaine Beasts." The emphasis here underlines the materiality of the text that was so radically missing from the first lines of the title. It seems to chide readers into recalling that loose secrets are difficult to hold, and books are ideal places to keep them organized for their own use and protected from the uninvited attentions of others.

The author makes his second bold appearance in this final subtitle, sending readers who needed a reminder back to the first line, where they would encounter the large black type attributing these secrets and the book itself to Albertus Magnus. He is not the only person recognized on the frontispiece: William Jaggard, the printer, is featured in larger type than the second authorial attribution, and his name anchors the page, with the date of printing centered beneath it to remind readers of the currency of this edition. "Printed for W. Iaggard" is separated from the final subtitle by an ornate woodblock design of interlocking cloverleaves that is almost exactly the same width as the last line of print.

Cornucopiae, Or divers secrets addresses more general human concerns on its frontispiece, but it presents its contents as immediately useful (Figure 4). The subtitle advertises "the rare secrets in Man, Beasts, Foules, Fishes, Trees, Plants, Stones, and such like," an announcement that reinforces the book's task and subjugates the natural world to human curiosity. It also guarantees the novelty of the information it presents, assuring readers that these secrets are "newly drawn out of diverse Latin authors into English" and have not been "before committed to be printed in English."

The frontispiece mentions the translator and the printer in type of equal size, suggesting that each played an integral role in bringing forth this book of secrets. The dominant image on the page, however, is the printers' device, which balances out the large print and capital letters used for the word *Cornu-*

copiae and warrants its own line. The alliance between the title word and the printer's device makes it apparent that the printer was the primary producer of this collection of natural knowledge, while the translator slips into the background, overshadowed even by the location of William Barley's print shop.

Crafting Readers of Secrets

An analysis of the reader's persona, as opposed to the desired reader characteristics developed above, begins with the letter to the reader. *The Book of Secrets of Albertus Magnus* begins with a note to the reader addressing the anonymous translator's decision to translate this book because of the positive reception it has had throughout Europe in other languages. The letter then continues to instruct the reader on the best way to make use of this book. "Wherefore use thou this book to mitigate and alacrate thy heavy and troublesome mind, as thou has been wont to do with the book commonly called the book of Fortune."[13] The letter to the reader in these editions of *Albertus Magnus* manages to make this text part of an international knowledge tradition, reminding readers of its success in Europe and its place in continental popular print, and part of the emerging English print market, particularly the part dominated by perpetual almanacs and addressing questions about natural knowledge, magic, and prophecy.

The title establishes it as part of a medieval knowledge tradition, one that includes Albert the Great, and it reinforces the book's desirability with its Latin version of the author's name and its attribution of the contents to a medieval friar and bishop who, even in the sixteenth century, had achieved a reputation for his knowledge of the natural world. The 1599 edition of this text, printed by William Jaggard, presents an elaborated version of this letter that reiterates its links with medieval knowledge, continental intellectual circles, and English popular print and makes two interesting additions. The first addresses the faithfulness of the translation: "Therefore I have in this translation omitted nothing, which therein [in the Italian, French, and Latin editions] is published: but thou shall find therein one later addition of the governance of the seven Planets in the nativities of children, which is worthy noting."[14] The provenance of this work is not made manifest, but it seems likely that it was borrowed from another book and author and tacked onto *Albertus Magnus* to refresh an established seller. The second addition follows the suggestion that readers consult this book for recreation and amusement and addresses the accuracy and utility of the secrets it reveals. "I refer thee to the trial of some

CORNVCOPIÆ,
Or diuers secrets:

Wherein is contained the rare secrets in Man, Beasts, Foules, Fishes, Trees, Plantes, Stones and such like, most pleasant and profitable, and not before committed to bee printed in English.

Newlie drawen out of diuers Latine Authors into English by Thomas Iohnson.

AT LONDON.
Printed for VVilliam Barley, and are to be sold at his shop at the vpper end of Gratious streete nere Leaden-Hall. 1 5 9 6.

Figure 4. Frontispiece of the *Cornucopiae, or diuers secrets,* London: Printed for William Barley, 1596. This image is reproduced with the permission of the Huntington Library, Art Collection, and Botanical Gardens in San Marino, California, where this book is held in the Rare Book Collection under the call number 16916. Further reproduction is prohibited without permission.

of his secrets, which as thou shalt find true in part or all, I leave to thine own report or commendation."[15]

The importance of this last sentence lies in its acknowledgement of the fact that readers valued this book for its recipes and information about the ways in which different natural objects could be used to manipulate the world around them, rather than just as a toy for their amusement. William Jaggard believed that readers planned to use the book, rather than just read it. He also gave them the chance to judge its contents. His willingness to grant authority to readers, to give them the chance to pass judgment on the quality of the secrets of the renowned Albertus Magnus, underlines the fact that he believed that the book was going to be understood according to its readers' needs and abilities. Under these conditions, the validity and utility of the secrets would always be subject to readers' perception, and readers would be the ultimate authorities. Even the pursuit of natural philosophy could come under readers' attack, particularly in the context of defining natural relationships and effects and making them the purview of acceptable intellectual inquiry.

The possibility of a highly flexible category called "natural" appears for the first time in the letter attributed to Albertus Magnus, which revolves around a justification of science and magic as equally powerful ways of understanding relationships in the natural world. It asserts that, according to Aristotle, "every science is of the kind of good things" then continues, "the science of Magic is not evil, for by the knowledge of it, evil may be eschewed and good followed." Neither magic nor science can be perfect, however, since both have the potential to be used for evil, which is defined as "beholding of causes and natural things."[16] This letter reinstates the early modern belief that magic was a natural knowledge system that took advantage of the relationships between things in order to benefit the performer. It also asserts a flexible definition of acceptable natural philosophical inquiry, which approves of the manipulation of all natural relationships, as long as they are not harmful. This definition of natural suits the assertion that readers planned to use this book and hoped to avoid traipsing into supernatural entanglements. It also establishes the reader as an intelligent, questioning voice at the beginning of the text, a role that will gain credibility after the books of virtues, which take a rather less generous approach to the task of laying bare the secrets of nature to satisfy readers' needs.

The books of virtues were structured to meet a specific set of readers' needs. These needs fall into several categories: The first is medical advice, such as curing or preventing sickness or healing wounds; the second is interpersonal success, including improving upon desirable personal characteristics, vanquishing enemies, and doing well in business; the third centers on getting, keeping, or verifying the loss of love; and the fourth, mystical accomplishments like seeing the future and preventing a woman from eating at your table.

While the books of herbs and beasts are presented with the admonition to restrict the manipulation of nature to astrologically advantageous times, the book of stones concludes with a stern warning to the reader to monitor his or her physical and moral state before performing any of the recipes contained in the book. The warning reads, "The maner of doynge these thynges, consisteth in this, that the bearer for a good effecte, be cleane from all pollution, or defylynge of the bodye."[17] Readers were told that they had to be physically clean, which would, according to Aristotelian notions of the polluting power of menstruation, have prevented menstruating women from using the information they had learned. The word "cleanliness" also has a moral aspect, equating sexual restraint with bodily purity.

The problem of satisfying readers' sexual desire and affections comes up more frequently in this book of virtues than in the other two. It is particularly concerned with discovering sexually wanton women who have either committed adultery or are attempting to enter marriage having already had sex. In fact, it devotes three of its forty-three entries to recipes that elucidate female sexual behavior, more than it devotes to any other topic.[18] The book begins with an entry entitled, "If thou wilt knowe whither thy wife is chaste, or no," offers a second set of directions headed, "If thou wylte knowe whither thy wife lieth with any other married man, or no," and concludes with an entry headed, "If thou wylt cure a vyrgen."[19] The third recipe appears to be different from the other two, in that it does not seem to provide instructions for policing a woman's sexuality. In fact, it does just that, since a maid claiming to be a virgin who was not cured by the entry's prescribed use of the stone could easily find her virginity in question. All these recipes seem to assume a male reader, since a woman would be likely to know her own sexual status, and the only information she could gain from these entries is an awareness of the tests that her father, husband, or fiancé could have in mind for her. While the recipes do not invite that kind of use, they certainly allow for it, and it is possible that female readers became very careful about how they slept and what they washed after they read *The Secrets of Albertus Magnus*.

The primary purpose of these recipes, for men to covertly test the sexual fidelity of their wives or the veracity of their daughters or future brides, is made evident by the sneaky tone of their directions. In the first recipe testing a wife's chastity, for example, the lodestone is to be put under the sleeping head of the woman in question, and her fidelity depends on whether she embraces her husband or falls out of bed. The second paragraph of this entry suggests that this test would have convicted a large number of women for infidelity, since it notes that a slightly more intensive use of the lodestone has the potential to make sleeping members of a household flee, leaving all their possessions behind.[20] Any stone that has that kind of dispersing power must have been capable of rolling a few innocent women out of their beds. The second recipe can also be performed without the woman's knowledge, as it involves handing her the stone Galeritis to wash and observing whether she immediately needs to urinate. The text of the heading and the entry conflicts, though, over the exact nature of the sexual information that can be gleaned from this test. The heading, as shown above, advertises a test for wives who may be having an affair with a married man, while the entry claims that the washing of Galeritis can only determine whether a woman is a virgin.[21] A close reading of one of the entries, the recipe for curing a virgin, reveals the ways in which these categories erode under the pressure placed on them by the heading.

"IF THOU WYLT CURE A VYRGYN.

Take the stone which is called Saunus, from the yle Sauna. It doth make firme, or consolidate the mynd of the bearer of it. And beyng bound to the hande of a woman, travayling with chylde, it letteth the byrth, and kepeth it in her bealye. Therefore, it is forbidden in suche a busyness, that this stone touche a woman."[22]

The first two lines ally the reader with the bearer of the stone, and thus subject her to its purported effect of consolidating the mind. The third line introduces the stone's capacity for stopping the delivery process if it is bound to the hand of a woman in labor, and the final sentence makes the reader responsible for making sure that this does not happen. Who, after all, were these readers? They cannot be only men who were interested in testing their wives' fidelity, since very few men would have had access to pregnant women in labor in sixteenth-century England. The care of the women and children and the management of labor and delivery fell to other women, particularly midwives, except in the most dire of circumstances when a surgeon might be summoned to extract a baby from a dying or newly dead mother. Midwives would have been most eager to prevent a long, complicated birth, if for no reason other than it would have reflected badly on their professional reputations, so the information about Saunus would have been useful to them. They would have been attracted to the entry by the heading as well, since the care of virgins, or at least young, unmarried women who were supposed to be virgins, also fell within their province.

This entry may be intended specifically for midwives or other women, but *The Secrets of Albertus Magnus* is far from a midwifery manual, and its readers were more likely to be men than midwives. If a male reader is assumed, this passage becomes much more challenging. The first thing to be called into question is the word "cure" in the heading: after all, what about an unspecified virgin needs to be cured? Men interested in "curing" virgins might have read on to the first two lines of the entry, which ensure for the bearer of Saunus a firm mind. Men who have decided to have sex with a virgin would need one, since they would know that they were increasing her chance of becoming pregnant and their own chances of becoming responsible for her and a baby, if they were found to be the father. They would have made their own connection between virgins and laboring women before they even arrived at the preventative effect of Saunus on delivery.

The admonition that concludes this book of virtues, which warns about the dangers of using any of the recipes when morally or physically unclean, suggests that there was some concern about the ways in which readers could make use of this book and natural philosophy to pursue bad ends, which would in turn threaten the acceptability of the discipline and this kind of book, since the text has already established that magic and the manipulation of nature are acceptable only as long as they result in good ends. The problem of defining good ends is left to the reader, who is also asked to govern his or her desires according to society's moral code. The place of morality in governing the pursuit and use of natural knowledge is most evident in the entries that address questions of sexuality and love, but the force of the stars in governing human behavior shapes the use of all the knowledge in this book.

Astrology and astrological theory are a dominant force behind the functioning of the natural world in *Albertus Magnus*. All four sixteenth-century editions conclude the Books of Beasts with an introduction to the movements of the planets and their effects on chronology, arguing that this information is crucial to implementing the other knowledge presented in the book. "And that all things which hath been said before, and also shall be said after, may be applied more easily, to the effects of their desire, which have not cunning of the stars."[23] Readers would have been moved to continue perusing this section, since it offered them the chance to expand their hold over and understanding of the natural world.

The discussion continues with a dissection of the sun's movement and how it relates to the hours of the day. It presents two ways of understanding time based on an assumption of equal and unequal hours. Only one sentence is dedicated to the simple concept of equal, or clock-based, hours, while two pages introduce and describe the astrologically based notion of unequal hours. The model, which is attributed to anonymous "Astrologians," is mathematically rigorous and presents the movement of the sun and other planets as logically predictable.[24] The text provides both the theory behind the mathematics and the calculations required to apply the theory, a necessary effort considering that it forms the basis of the astrological system that assigns each hour and day to the dominion of a planet and determines the behaviors of plants, minerals, and animals, alone and in combination.[25]

This section proves that the stars have a consistent and predictable effect on natural objects, and that any attempts to manipulate or understand the natural world must be performed in accordance with astrological principles. The decision to position astrology in the center of the text and use it as a bridge between the similar first three books of virtues and the slightly different final book of marvels suggests that *Albertus Magnus* was meant to be read as a single volume united by theoretical approach and a coherent set of expectations for the ways in which the natural world should behave. William Jaggard's decision to expand the astronomical section by adding an entirely new set of material to his edition of this book points to the importance of this kind of knowledge for understanding and making use of the material in this text, and it attests to astrology's primacy in early modern models for thinking about the world. The expanded section on astrology simply makes this theoretical system both immediate and accessible to readers applying it to the nature, appearance, and bodies of children, even while it reasserts the position of stellar influences within the category of natural forces.

The astrological treatises testify to readers' intentions to make use of this text and their desire to maximize their knowledge about the factors affecting natural objects, but the presentation of this theoretical system simply functions to satisfy the demands of the same invisible reader who wanted to inspire love. The specific desires of the reader are relatively well delineated by the end of the astrological section, but their role in judging the accuracy and utility of natural knowledge and their relationship to experience is best demonstrated in the book of marvels. This book carries the considerable task of expanding the category of natural to include apparently wondrous occurrences by extending the definition of natural in the books of virtues to its fullest potential, while it defines the ways in which readers' experience can interact with the recorded experiences of authorities.

The lodestone appears here as evidence of the efficacy of hidden forces and as an illustration of their role in determining the structure of the natural world. "For although we know not a manifest reason wherefore the Lodestone draws to it iron, notwithstanding experience does manifest it so that no man may deny it."[26] It also functions to remind readers that their own experiences provide a valuable tool by which to make judgments about the quality of the knowledge they encounter. But in the end, the experience this section of the book privileges is that which verifies recorded knowledge, and no advice is given for how to interpret a gap between personal experience and textual authority. The segment on the lodestone concludes, "So marvelous things are declared of philosophers to be in things by experience, which no man ought to deny."[27]

The place of experience and the ubiquity of systems of sympathy in the natural world is more firmly defined. The text reiterates the belief that female palm trees bent down and opened their leaves to male palm trees in order to be inseminated. This was considered an exceptional display of sympathy in action, since the palm trees did not have to touch in order for the female to be inseminated. The example is, however, challenging because the force drawing the two together is invisible, and it is difficult to prove that insemination occurs only under these circumstances. The book goes on to record the details of artificial insemination in palm trees, as practiced by the ancients based on their observations of this occult transmission of virtues. "Reddit ergo erecta, supra seipsam quasi adepta sit Masculo per continuationem fumis Virtutem masculi."[28] This sentence indicates the way in which occult insemination was understood, highlighting the importance of invisible "vapors" that carried the male seed to the prostrated female tree. It also indicates the importance of what may have been a general agricultural practice in shaping natural philosophical knowledge. This conclusion is supported by the sentence that follows the Latin description of the palm trees, which rescues natural philosophical authorities from ignominy: "many of the ancient authors, hath shewed marvaylous thinges, received now of the common people and taken for a truth."[29]

The Natural World According to Albertus Magnus

The printers of *Albertus Magnus* and *Cornucopiae* integrated their attempts to produce a useful and believable map of the natural world into the text. One of the strategies common to all the printers was to introduce new information, like Latin or Greek names for each entry, which helped to guarantee the appearance of secrecy, alongside the things each object could do, which made them interesting to readers. The first three books of *Albertus Magnus* present in turn the secrets of herbs, stones, and beasts, and they begin with a table of contents for the entries that includes the name of each natural object in Latin and English. The contents follow this map and enlarge upon it by presenting an often creative version of the object's name in Greek and Chaldean as well as repeating the Latin and English names, then telling where the object might be found, its notable characteristics, and how it could be used by readers. This presentation style accomplishes at least three things: it advertises the exotic and learned nature of the knowledge it contains, it offers new terminology in relation to vocabulary readers would recognize, and it emphasizes the utility of the information it presents.

The first entry of the book of herbs, for example, begins with a recitation of Chaldean, Latin, English, and Greek names for marigold and a natural explanation for the origin of the Latin and English names based on the flower's orientation to the sun.[30] The array of foreign languages in the entry reminds readers of the academic roots of the source, and it emphasizes the exotic quality of the knowledge the entry makes available to readers, wrapping the familiar flower in foreign dress. The entry then continues "The vertue of this herbe is marvelous, for if it be gathered, the Sunne being in the sygne of Leo, in August, and be wrapped in the leafe of a Laurell or bay tree and a wolfs tooth be added therto, no man shall bee able to have a word to speake against the bearer thereof, but wordes of peace."[31] The rationale for adopting this presentation strategy, in which new knowledge is wrapped around familiar natural objects and addressed to readers' needs, is logical. Each entry reiterates the primary goal of the text, which aims to subjugate the natural world to human manipulation, and each emphasizes the natural relationships that can be manipulated to meet readers' needs. The entries also demonstrate the importance of considering natural objects in relationship to each other in order to determine their place or places in the natural world. That goal is furthered by the slippage that occurs within the contents of each book.

It is true that each of the main entries in the first three books in *The Secrets of Albertus Magnus* consists of natural objects categorized according to their essence—there are no stones in the table of contents for the book of herbs. Within each entry, however, is included all the herbs, stones, or beasts whose interrelationships lend the best advantage to readers in their efforts to manipulate nature. This results in a multitude of types of natural objects in each book of virtues, and they present these objects in the context of their relationships to the focus of the entry. The entry on marigold, for example, lists bay leaves and wolf teeth as the other two ingredients required to ensure that a man will have only peaceful encounters.

Neither the bay plant nor the wolf warrants its own entry in the book of herbs or the book of beasts, though both are mentioned in the book of marvels. Their depictions in this last context, which aim to reiterate the boundaries of the category "natural" and demonstrate its potential, are quite different from their images in the marigold entry. The wolf, which appears in the book of herbs only as a body part in relationship to marigold and bay, is lauded for his ferocity in the book of marvels and for the effect of that ferocity on men. The entire wolf is marvelous because "it is said, if the Woolfe see a man, and the man see not him, the man is astonished and feareth, and is hoarse."[32] The image of the fierce predator in the book of marvels is fleshier and more frightening than the disembodied tooth. Both these wolves, however, appear in relation to altering human speech, indicating that the wolf, either whole or in parts, has significant power over the human tongue.

The two images of wolves build upon each other to produce a coherent image of the animal, but this is not always the case. The sixth entry in the book of herbs is devoted to pennyroyal and reads "Take this herb and mix it with the stone, found in the nest of the bird called a Lapwing or black Plover, and rub the belly of any beast and it shall be with birth, and have a young one, very black in the own kind."[33] This entry is notable for its depiction of the black plover, which does have its own entry in the book of beasts, and for its representation of systems of sympathy. According to the relationships delineated in this entry, the stone found in the plover's nest either lends or adopts the bird's characteristic blackness and is then capable of passing it on to any animal it contacts.

The plover's entry includes mention of neither the bird's blackness nor the stone that can be found in its nest. Instead, it emphasizes the power that can be gained by carrying the bird's eyes, which ensure physical size and passive enemies according to where they are carried, and the bird's head, which protects its bearer from being fooled by any merchant.[34] The black plover appears two more times in the book of marvels. In a long line of entries delineating the effects of animals and their parts on humans, appears "Tabariences says if the tongue of the Lapwing or black plover be hanged upon a wall *Oblivionem reddit eum memorem et alienationes*" and, two pages later, "And if the heart, eye, or brain of a Lapwing or black Plover be hanged upon a man's neck, it is profitable against forgetfulness and sharpens man's understanding."[35] The four images of plovers presented throughout the book are unified through systems of sympathy, the primary model of natural change forwarded in *Albertus Magnus*. In this case, the unifying characteristic is that their different parts can have significant impact on people's brains, either by lending cleverness or improving or obliterating memory. Blackness and its method of transmission, which dominates the first mention of the plover, never appears again.

The combination of these images produces a complex plover complete with all its relationships to and effects upon other natural objects. Each image exists on its own as well, resulting in the possibility of readers simultaneously entertaining four singular plovers and a fifth conglomerate one, as determined by their needs and interests. *The Secrets of Albertus Magnus* lends itself to two kinds of use because of this presentation style: It can provide the basis for creating an image of the natural world characterized by the relationships between objects and their vulnerability to human manipulation, and it can be used to pursue answers to personal questions. The tables of contents that head each of the books of virtues supports the viability of using these books to answer specific questions or satisfy specific needs, while the structure of each book indicates that those questions will be answered through the manipulation of the connections between or among natural objects. The flow between the books, and particularly the development of the definition of nature and the place of readers and experience that culminates in the book of marvels, permits readers to produce personalized images of the natural world that derive their power and appeal from the potential for humans to mimic natural processes and cause desirable changes.

Expanding and Systematizing the Natural in Cornucopiae

The task of defining the scope of "natural" is not solely the province of *Albertus Magnus*. *The Cornucopiae, Or divers secrets* announces its intention to expand the definition of "natural" by maximizing reader accessibility and perceptions of utility. It begins with the pronouncement "Many are the wonders and marvels in this world, and almost incredible were it not that experience teaches the contrary."[36] Experience guarantees the value and utility of the knowledge in this book, and it comes in two types. The first comes from the original author of each entry and, by association, the compiler. This kind of experience can be exotic, in that it is often demonstrated with examples of animals, plants, stones, or interactions that are not native to England or even Europe. In these cases, the author's experience has to justify the presence of the entry in the text, and it has to be sufficiently authoritative to guarantee the entry's validity.

The lines following the book's introductory statement are an excellent example of this strategy for creating and making use of authorial experience. It reads "for who could be persuaded to believe that the Owstridge could eat or devour cold and hard Iron, or that hot burning Iron could not hurt her stomach."[37] The fantastic nature of ostriches' eating habits verifies the need for believable authorial experience, since the vast majority of readers would not have seen an ostrich or even been capable of clearly imagining one. They would likely, however, have been familiar with the fact that remarkably few animals eat iron, hot or cold, and remain unscathed. The text addresses the difficulty associated with verifying an experience that so drastically opposes most people's conceptions of the ways that nature works. One of the strategies it uses to create legitimate experience is repetition, suggesting that no one could be expected to believe a single fantastic report but that multiple reports lend credibility to even unbelievable claims.

Readers are expected to accept these multiple narrations of incredible experiences as reliable, and they are given a context in which marvelous and exotic things are the same as marvelous and familiar ones. In order for readers to make this leap, they first have to believe that their experience with nature does not conflict with the exotic experiences of the authorities in the text. They must also believe that their experience and needs, not just the experiences and needs of various authorities, are valid and a worthy center around which the text can coalesce. The book supports this contention in very specific ways. First, readers' experience is recognized as the force that shapes their reasons for reading and methods for using the information in the book. Second, readers are granted expert status in determining and seeking remedy for perceived personal lacks. Entries that offer remedies for insufficient love or insufficient bravery, for example, recognize the readers' ability to identify lack in their lives.[38] Their expertise derives from their life experience, a part of the natural world that remains opaque to even the most authoritative experts, and it dependably allows them to recognize their own limitations.

The second type of authority commanded by readers in *Cornucopiae, Or divers secrets* is the chance to use their own bodies to demonstrate relationships in the natural world. The best example of this is the section in which readers are invited to test the claims about relative proportions of the body. It begins, "Wonderful are the portraitures and proportions of men. Let a man stand upright and hold up his hands over his head, then is there his cubit from the top of his head to the long fingers ende," then continues to list different ways of measuring the body according to the relative length of its parts.[39] In this case, the readers bring to the text the means of testing its claims, and in so doing, their physical experiences become validation for the text.

The book produces two coexisting, equally valid models of experience, a subjective, individualized one and a cooperative, generalized one, each of which contributes to a specific model of the world. The first one contributes to models grounded in readers' needs and perceptions, while the second integrates those needs with the authors' broader claims about relationships in the natural world. The value placed upon readers' experience is very similar in both *The Book of Secrets of Albertus Magnus* and *Cornucopiae, Or divers secrets*. In both sources, experience is used to expand commonly held definitions of the natural to include wonders, such as iron-eating ostriches, that had previously been inexplicable and outside the realm of the natural. In a time when so many inexpensive books were trumpeting breaches in the natural order, such as three-headed calves, as the results of divine or demonic interference with the material world, books of secrets are exceptional and correspond better with contemporary ideas in university-based natural philosophy. The late sixteenth century saw the beginnings of an effort to expand the category of natural and minimize the place of preternatural and supernatural in human experience with the world, and these books are structured to support this movement.[40] Books of secrets stand out among other natural philosophy texts, however, for providing readers with natural worlds that correspond to their needs as well as the larger laws governing change in the material world. These books allow readers to construct personal and specific as well as general images of nature, and those images allow the reader to change position in relation to the larger world. The personal model obviously centers on the reader's needs and desires, while the more general one can be manipulated by the reader but also makes her subject to the same rules governing the things she is reading about. Taken together, these two models lend insight into the ways in which secrets stayed secret—since they were subject to readers' interpretation and personal requirements, their meaning was fluid and their revelation always potentially new.

Sympathy, Antipathy, and Secrecy in Albertus Magnus and Cornucopiae

The problem posed by keeping printed knowledge secret is directly addressed by the structures of both books and is intertwined with the types of worlds they encouraged readers to build. *The Secrets of Albertus Magnus*, for example, reviews extant theoretical systems in order to forward one that favors a combination of inherited knowledge and personal experience aimed at interpreting relationships among natural objects. The resolution of this system takes place in the book of marvels, where the text explores the fundamental principles of Aristotelian natural philosophy, systems of sympathy, astrology, and alchemy in order to arrive at the conclusion that every natural object has its own constitution and is linked to others through ties of sympathy and antipathy that comprise the relationships that form the means of creating, predicting, and manipulating action in the natural world.

The book begins with a restatement of the fundamental nature of sympathetic action in determining the behavior of natural objects: "After it was known of philosophers that all kinds of things move and incline to themselves, because an active and rational virtue is in them, which they guide, and move as well to themselves as to others, as fire moves to fire &c."[41] By this point in the text, the assertion of sympathy as an integral force in natural relationships is familiar, since it appears consistently throughout each of the books of virtues, beginning with the multiple effects of the seventh entry in the book of herbs, hound's tongue, on dogs.[42] This version of sympathetic interaction depends upon the doctrine of signatures, which makes it likely that a plant that resembles a dog's tongue will be able to affect actual dogs' tongues.

The book of marvels does not rest with this elemental version of sympathetic action, but goes on to assert the importance of association in determining systems of sympathy in natural philosophy. It reiterates the power of association, a message made clear first in the sixth entry of the book of herbs with the effect of the black plover on the offspring of other animals: "And Avicenna says, when a thing stands long in salt, it is salt, and if any thing stands in a stinking place, it is made stinking."[43] It then underlines the importance of determining the constitution of each natural object and whether its defining characteristics are inherent or acquired in order to understand its effects on other objects. It claims this as a fundamental unifying characteristic of natural philosophy, popular understandings of nature, humoral medicine, and alchemy: "It is not hidden to the people that every like helps strengthen his like, and loves, moves, and embraces it. And Physicians have said, and verified, that the liver helps the liver in their writings, and every member helps his like. And the turners of one metal into another called Alchemists know that by manifest truth, how like natures secretly enter and rejoice of his like."[44] Sympathy and antipathy between objects determine the composition of the natural world, and they act as the levers by which readers could mold it according to their needs and desires.

The theoretical section of the text concludes with an assertion that makes the book a viable collection of methods for influencing the world, rather than a random assortment of interesting facts: "And it is manifest to all men, that man is the end of all natural things and that all natural things are by him and he overcomes all things."[45] This statement of human superiority makes it possible for readers, even those whose contact with nature has been consistently as its victims, to believe that they can take control of the world around them. It also makes evident the book's function, which is to help readers capitalize on their advantages through specific manipulations of natural relationships. Finally it helps them to reimagine the natural world as flexible and capable of meeting their needs. None of the theories that contributed to this conclusion would have been entirely foreign to early modern readers; as the book notes, everyone knew that likes attracted. A concise and affordable explication of these theories would, however, have been difficult to find in English before the first editions of this book.

The knowledge conveyed in the book of marvels empowers readers to take control of their worlds, and it completes the text's project of making readers' worlds subject to their needs and desires. The knowledge in the section is secret, even if some of it is widely known, because it has never before been presented in this way—in English, for such a low price. It remains secret, even after it is widely available, because it can be used only by certain kinds of readers: those who are willing to heed the repeated warnings to be cautious with the manipulation of nature, to make use of their knowledge only after they have learned it thoroughly, and only at the appropriate astrological moments that permit each relationship to occur as described. Warnings to readers on how to properly use their new knowledge dot the text, appearing at the beginning and the end of each book of virtues and throughout the theoretical section of the book of marvels. The effect of this repetition is to create an image of the desired reader and practitioner of secrets that features diligence, thoughtfulness, and studiousness as his and her most important characteristics.[46] It also favors selfless and open-minded readers who pursue the secrets of nature as part of a larger understanding of natural philosophy and who are excited by new developments.

The general importance of sympathy in governing relationships may not be "secret and hid from the people," but it is useful only to people who are willing and able to obey the order to "turn thou not away the eyes of thy mind" when its full impact is revealed in the texts of natural philosophers like Aristotle.[47] The ideal reader of secrets appears in opposition to people who have disregarded or disparaged the theories of natural philosophers to protect their own status as experts: "And they that understood not the marvelousness and how that might be, did despise and cast away all things, in which the labor and wit of philosophers was, whose intent and labor was their own praise in their posterity."[48] These people make observations and cite experiences that cannot be trusted, since their personal investment in specific theoretical systems blinds them. Good readers, on the other hand, are open to accepting theories and experiences authorized by experts, and their ability to decipher true from false observations renders them capable of understanding and using the secrets of nature.

The Cornucopiae, Or divers secrets shares the emphasis on sympathy, antipathy, and interrelationships in *The Secrets of Albertus Magnus* as the primary causes of action in the natural world, but it has a very different approach to introducing its system of natural knowledge. The text begins, "Many are the wonders and marvels in this world" and ends with an assertion of the importance of observation in determining the validity of those wonders. This move from reported wonder to observed occurrence results in an alliance of wonders and natural relationships. Once the testimony of multiple witnesses renders something believable, it leaves the realm of the suspect and enters the realm of the observed, which includes a range of things that may, upon first glance, seem difficult to believe. According to this text, wonders and marvels that have been recorded by legitimate witnesses simply form part of the plethora of natural things to be experienced.[49]

Once the categories of the marvelous and the useful are collapsed into a single entity that provides insight into the organizing structure of the natural world, it becomes apparent that insight into underlying patterns will give readers the ability to predict and even control their environments. It also makes it possible to unfold the underlying patterns of the natural world according to each entry's relationship to another, rather than making each entry answer to the perceived needs of the reader. This book moves from entry to entry, carrying the reader through a chain of connections that reflects the linkages between and among elements in the natural world. It requires no explanation, since the connections between each entry and the one following it are evident, and it requires no headings, since the transitions between groups of entries are determined by their relation to each other. Finally, it requires no theoretical basis, since the logic governing the ordering of entries is entirely internal.

The first entry in the text reads "There is a little fish called Echines, which cleaving to the keele or maste of the Ship, so retaine the shippe that no violence of winde or weather can remove it."[50] It continues with two more entries that defy expectations about the relationships between objects in the natural world, including salamanders and piranha, which are said to live unharmed in fire, and satyres, which have the heads and intellects of men and the bodies of goats. These three examples disrupt readers' notions about the way the world should work and introduce them to the governing logic of the book, which moves from one natural object to another according to links determined by sympathy or association. Like the knowledge in *The Secrets of Albertus Magnus,* the information here has the potential to have already been revealed, but the systems of sympathy that govern the text and provide the basis for understanding the ways in which the natural world operates would have been secret, not because systems of sympathy were unknown, but because they had not previously been so demonstrated.

Structuring for Secrecy in Albertus Magnus and Cornucopiae

The struggle to make clearly printed knowledge secret contributes to the structure of *Cornucopiae* and *Albertus Magnus.* The first refuses to categorize objects at all, so that their various relationships with each other multiply and contribute to the idea of secrecy. *Albertus Magnus,* on the other hand, places objects in categories to demonstrate the multiple ways in which many things can operate. As demonstrated above, *Albertus Magnus* offers multiple images of black plovers and wolves, both of which appear in various places throughout the text. While their images are not necessarily consistent among the four books, they can be combined to produce an image of a natural object characterized by its powerful relationships with a range of other objects. Plants, stones, or beasts that merit multiple mentions stand out in the text as more significant than other natural objects and therefore more central to readers' plans to change the natural world. They also extend the reach of the categories that shape the book, since they suggest that none of the categories exists independently, and that each gains its power and identity through connections with the others. Despite this structure, the texts remain reader-centered, and its composition depends on linked signifiers.

The image of the eagle in this book provides an excellent example of the power and influence of repetition on the ways in which the book structures its depiction of the natural world. The plant aquilaris, for example, makes its first appearance in *The Book of Secrets of Albertus Magnus* as the fourth entry in the first book, which elucidates the various medical and magical effects of herbs. The entry reads, "The fourth herbe named Aquilaris of the Chaldies, because it springeth in the tyme in which the Eagles buylde their nests."[51] Aquilaris is replaced in the second book, which concentrates on the various characteristics and effects created by stones, by Aquileus, an object again defined by its relationship to eagles. It reads, "Take the stone which is called Echites, and it is called of some Aquileus because the Egles put these in theyr nestes."[52] Eagles appear several times in the fourth book, a collection of the world's marvels, as the terror of the avian world: "all flying Byrdes flye the Eagle" and "when the feathers of Eagles bee put with the fethers of other fowles, they burne and mortify them: for as hee overcommeth in hys lyfe al Birdes and ruleth over them, so the feathers of Eagles ar deadly to all feathers."[53] From aquilaris to aquileus to the power of eagles, this chain of linked signifiers moves the reader back and forth in the text, prompting recollections and backtrackings that reward the reader with a rich picture of the interconnections between natural objects that becomes integral to their image of the natural world. The reader remains at the center of that world, however, since it is structured by his or her needs, and it is constructed from his or her experiences with and images of specific natural objects.

The impact of the repetition is to remind readers that the world rebels against systematic containment but responds to specific manipulation, and that their efforts to do so must correspond to the flexibility of natural objects rather than the strictures of artificial categories. Repetition in *Cornucopiae* signals the entries that demonstrate fundamental principles governing action in the natural world, specifically what constitutes natural action, that every natural object has its own characteristics and relationships with other natural objects, and that sympathy and antipathy are the two driving forces behind relationships in the natural world. The magnet, which appears four times in the early pages of the text, demonstrates the function of repetition. It is particularly useful for accessing the place of repetition in this text, because the medieval world had struggled to explain its operation in natural terms.[54] It appears for the first time as the fourth entry in the text, acting as a transition from exotic examples of natural wonders to the recitation of the characteristics of familiar stones. "The Loadstone has virtue to draw iron to it: yet if you hold a Diamond by him, that virtue will be taken away so long as the Adamant is by him."[55] Here this familiar, if contested, image of natural force reassures readers that salamanders, piranha, satyrs, and small but remarkably strong fish are as natural as, even if more wonderful than, the beneficial effects caused by the stone found in an eagle's nest, which can ease labor and delivery, and the jasper stone, which can stop bleeding.

The next image of the magnet demonstrates the importance of contact in understanding sympathetic relationships, particularly those that create new things or lend the characteristics of one natural object to another. It reads, "The Loadstone doth not only draw iron to it, but also makes that iron draw other iron to it, if the Loadstone be rubbed therewith."[56] The magnet appears again on the next page in a list of stones and their virtues, and the section operates as a transition to a list of the questionably natural virtues of various plants, like the fact that male and female palm trees embrace.[57] Its final appearance marks the introduction of the last governing force in natural relationships, antipathy. This time, the diamond is the focus of the entry and the magnet's loss of attractiveness serves to reflect the power of systems of antipathy.[58]

Repetition serves to reinforce the principles shaping the natural worlds in *Cornucopiae*, but a second strategy is used to move readers through the entries so that they can experience the effects of that repetition. In this book, each entry is linked to the next through a shared characteristic, such as location, means of action, type of object, or effect. For example, an entry on the magnet's ability to lend its iron-attracting properties to iron that is rubbed on it is followed by one on the impudence-lending characteristics of a harlot's dress, an entry on the same characteristic contained in a harlot's mirror, and then an entry on the depressive effects of coffin drapery worn later as apparel.

> The Loadstone doth not onelie draw iron to it, but also maketh that iron to drawe other iron to it, if the Loadstone be rubbed therewith. It is supposed that in like manner the smock or other apparel of a strumpet being worn of others, giveth a certaine impudencie and shameles boldness to those parties. Even so if a woman behold her selfe often in the glasse wherein a whore hath accustomed to look in, it maketh her not onely impudent bold but also the more prompt to further offending. Also a blacke cloth which hath beene used over the Coffin of dead follkes bringeth a certain kinde of sadness or melancholy to them that weare it in apparrell.[59]

These entries are linked by their effects, since each object has the ability to lend to others either its own characteristics or those of its user. The structure of this subsection also represents a movement toward generalization of these effects: while the lodestone can affect only iron, smocks and mirrors affect all female users, and funeral draperies affect anyone who later wears them.

The movement across entries highlights the presence of tightly linked, short chains of objects that demonstrate multiple examples of a single governing principle of natural action. The geographical section, which comes at the end of the book, is an excellent example of the ways in which linked entries reflect the effects of what may be the most important factor governing the natural world: that each natural thing has an innate set of properties that affect other objects around it. The first five entries read, "Above other countries, Spain, Iberia, Dalmatia, Tolosa, India, and the Ethiopian Islands are most fruitful of gold. . . . Corinthus, Caristos, Dodoua, abound in Brass. Above all other places, England, Italy, Thracia, and Calabria excel in multitudes of beasts and cattle. Africa and Arcadia in plenty of asses. For plenty of Trees, the mount Atlas and Caucasus, the fortunate Islands, India, Cirene, and Hiccinia."[60] This section expands the category of innate characteristics to include the natural bounty to be found in various lands. The next set of entries describes specific locations that harbor remarkable natural gifts, and these examples operate as the broadest statement of innate characteristics. These two sets of linked entries work in different directions, with those demonstrating the importance of contact becoming more general and the geographic entries more specific. This apparent antagonism does not disrupt the way the book moves, but instead suggests that it can work to build worlds at the level of each subchain or at the level of the entire text. The trick is moving readers within each chain and among chains in order to produce coherent depictions of a governable natural world.

The flow between and across entries is critical to the world-building project, since it moves readers from a general understanding of a single effect to a rich, multidimensional comprehension of the way that effect functions in different contexts. Transitional entries, those that pull the reader from an elaboration of one characteristic in multiple settings to a discussion of another characteristic, are fundamental to the book's flow. A chain of entries demonstrating the effects of plants, stones, animals, and people on each other reiterates the importance of sympathy and antipathy, even in relationships that cross elementary categories. The section begins with the logical example of the lamb fearing the wolf, then moves on to the hatred of the olive tree for whoredom, the lion for firebrands, and the wolf for thrown stones before returning to animals' enmity for each other.[61]

This chain of entries is important for its demonstration of the depths to which systems of sympathy and antipathy govern action in the natural world, and the ways in which they determine the way the world works. The themes of attraction and opposition repeat throughout the book, appearing in multiple entries in relation to a range of natural items. Objects like the lodestone can act as a successful transition multiple times because they demonstrate more than one governing principle, but other transition points appear only once and function to give a definitive example of a single natural property. This is best demonstrated by two examples that appear approximately three-quarters of the way through the book and reflect the necessity of multiple observers and experimentation in proving a questionably natural event.

The first of these two entries notes that upon Midsummer Eve a coal that will prevent plague can be found at the root of every plantain or mugwort plant. It goes on to assert the author's personal experience with both the truth of the coal's appearance and its protective aspects. "I dare be bold to note it for truth, for that I my self have found it diverse times in the presence of many at that hour, and having sought for the same at other times, it is not to be found: ... yea, and I dare assure you that I never knew any that ever carried it about them that ever had the plague or was troubled with any kind of ague."[62] The second entry reinforces the importance of experience in verifying the effects of mugwort on ale or beer, which protects it from spoiling in the summer. Both these entries are framed as extraordinary, but the use of experience and even experimentation to verify their existence and effects recovers them for the realm of the natural, where they ably demonstrate the fantastic limits of the impact of relationships between objects in the natural world.

Repetition is the dominant structural strategy in printers' depictions of the natural world in *Cornucopiae*, but *Albertus Magnus* contains a second strategy that reflects the position of the reader in its depiction of the natural world. The repetitions of objects and relationships and the consistently eroded and malleable categories in *Albertus Magnus* demonstrate the reader's position as the textual center, rather than the absence of any center at all. The books are structured to function entirely according to readers' needs and to instruct their perceptions of the natural world, its objects, and relationships. The reader, who appears in person only in the introductory letter to *Albertus Magnus* and as the *you* in phrases like "If you would move love" in the books of virtues is the center around which all the relationships in the natural world coalesce.

Conclusion

The Book of Secrets of Albertus Magnus and *Cornucopiae, Or divers secrets* represent the portion of the inexpensive print market devoted to revealing knowledge about and recounting various phenomena in the natural world. They were among the first texts of this sort available to an audience encompassing a variety of educational and socioeconomic levels. They were also critical to the movement that introduced inexpensive textual support for classical and medieval knowledge traditions that were already circulating in academic culture and, in nontextual form, as part of popular knowledge about the natural world. Both books cite recognizable authorities such as Aristotle and Pliny as the sources for their collections, and they note the role of translation and the importance of compilation in their creation. They also make significant statements about the definition of natural action, the place of humans in the natural world, and the relative positions of inherited knowledge and experience in relationship to natural philosophical authority. Finally, they illustrate the role of broad compendia of secrets and point to the ways in which books like these could be read in early modern England.

These books shared the intention of making the natural world appear vulnerable to readers' manipulation, and they shared the associated task of instructing readers on the best ways to bend nature to their wills. In order to accomplish the first of these goals, they created images of a world that was composed of systems of sympathy and antipathy between natural objects defined by the primary characteristics inherent in each object. They presented these relationships as the source of change and movement in the natural world and offered readers multiple examples of the ways in which those relationships could be manipulated to produce personally beneficial results. They grounded this knowledge in classical natural philosophical traditions and attributed it to recognizable authorities whose work had not previously been available in English, in order to assure readers that they were in fact purchasing secrets about the natural world.

In order to maintain the secrecy of the knowledge they were sharing, they created specific images of readers of secrets characterized by open minds, a willingness to subjugate personal experience with nature to the testimony of authorities, pure intentions for learning the inner workings of the natural world, and the desire to understand the place of educated people within that world. Some of the information contained in these books is targeted at the less scholarly and more human side of readers, who were expected to seek control of the natural world in order to succeed in business dealings, vanquish enemies, and attract and maintain the love of another person. The rest of the knowledge attempts to satisfy human curiosity about the contents and workings of the natural world. Both books insist on a broad definition of the natural that includes occult forces that were difficult to explain in natural terms. They were able to do this because their intent was to make available to readers a broad variety of natural phenomena, rather than to explain each of them. This is not to suggest that these books are devoid of explanation, rather to say that they integrate the explanation of wondrous occurrences into their presentation of natural relationships, so that systems of sympathy and antipathy, even those that are difficult to detect, constitute the most significant causes of natural action.

This chapter has interrogated the ways in which the structure of two general books of secrets contributed to, and in some senses determined, the ways in which readers used them. It argues that both *The Secrets of Albertus Magnus* and *Cornucopiae, Or divers secrets* use strategies of repetition to underline the multiple positions occupied by some natural objects, and that those objects in particular would have become signifiers of the ways in which natural relationships influenced the fabric of the natural world. It asserts that both books invite readers to place their own experiences alongside those of authorities as long as they do not conflict, and that they place readers at the center of the worlds they built by gesturing toward the specific character of readers' desires for control over nature. Readers could construct both personal and general models of the natural world from these books, but both worlds would have shared the characteristic of vulnerability to human manipulation.

The next two chapters will move on from the readers created within books of secrets to the impact of books of secrets on their readers. Chapter 4 will examine the books of secrets that were produced for specifically female audiences to analyze the secrets deemed appropriate for women and the limits applied to their ability to manipulate the natural world, as well as the definition of femininity connected to possessing and wielding knowledge of natural secrets. It will interrogate the image of women produced by these books and the ways in which that image could have been appropriated by female readers and influenced their relationships to and efforts to change the natural world. Chapter 5 will examine constructions of masculinity as they were presented in Gervase Markham's book on the keeping and training of horses. It will examine the class divisions produced within that book that clearly allocated some responsibilities for the secrets of nature, particularly the secrets of caring for horses' needs and health, to lower-class and usually young boys. It will also contend that the image of elite masculinity produced by the book includes mastering the ability to bend nature to your will, at least in the sense of breeding and training highly successful racing and performance horses.

CHAPTER FOUR

Secrets Gendered: Femininity and Feminine Knowledge in Books of Secrets

In 1573 John Partridge and Richard Jones became the first English Renaissance team of a compiler and printer to create a book of secrets intended for a female audience.[1] *The Treasurie of commodious Conceits, and hidden Secrets* appears to be a rather staid combination of recipes for dinners, desserts, and herbal medicines, but the dedication letter to Richard Wistow, a gentleman and assistant in the Company of the Barbers and Surgeons, the letters from Thomas Curtesye and Thomas Blank who are both identified as gentlemen, praising and defending the author, suggest that creating this book for a general audience was considered risky.[2]

The risk cannot have been understood solely in financial terms, or the book would probably not have been printed at all. This book, instead, marks a balance between the compiler and printer's perception of what was appropriate for a female audience, their ideas about the desires of that audience for "secret" knowledge, and the larger question of what made knowledge secret, whether secret knowledge was inherently dangerous, and how secrets could be safely told. Furthermore, it points to the position of secret knowledge in the construction of femininity, particularly as it was understood to be shared by women who could read—therefore, a smaller percentage of the total population of readers than the other books addressed. In 1588 those questions contributed to the creation of another book of secrets directed to female readers.

The Widowes treasure has been attributed to John Partridge and was printed by Edward Allde for Edward White. It is a largely medical and veterinary text with some cookery mixed in, and it expands the category of knowledge that women could appropriately master. It is prefaced only by an anonymous letter to the reader either by Partridge or Allde attributing the recipes to a book created especially for a gentlewoman that fell, entirely by accident, into Partridge's hands and that he believed to be so useful that it ought to be made available to everyone. This type of letter is very common as a heading in cheap-print editions of natural knowledge, especially those including information and recipes gathered from a variety of sources and packaged as elite and protected. It helped to promote the secrecy of the knowledge and protect the compiler from accusations of greedily profiting from secret knowledge. In this case, it seems likely that it also legitimized selling this kind of knowledge to female readers, since it was supposed to have been taken from a book that was created specifically for an elite female.

In 1596 Richard Jones produced *The Treasurie of hidden Secrets*, which combined the 1573 Partridge text with another cookery book printed by Jones called *The good huswives handmaid for the kitchin*.[3] A signed letter from the printer and a poem from the anonymous author introduce the text and attribute its creation to the request of an(other) anonymous gentlewoman whose ideas about its widespread utility brought it to light. This book lacks the veterinary emphasis of the previous one, but it provides a broad arena of natural knowledge that women could be expected to master and manipulate to their needs.

This chapter will analyze the structure and content of all three texts to place them within the genre of cheap print intended for general audiences and the subgenre of inexpensive books for women. It will continue to emphasize the construction of authority established in chapters 2 and 3 to concentrate on the ways in which it was created in these books for both men and women. It will extend the emphasis on the natural worlds that books of secrets allowed readers to create, begun in chapter 3, and focus upon the multiple natural worlds available in each book and the methods the books offer readers for manipulating those worlds. It will dissect the ways in which those methods are, and sometimes surprisingly are not, gendered. It will also trace the relationship in the books between representations of femininity, female readers, and secret knowledge. Finally, it will argue that these books were powerful, not just because they offered strategies to women for controlling the natural world and ways of being feminine that included controlling nature, but because in doing so they bridged intellectual and social communities and threatened the divides between accepted structuring binaries such as male and female, common and reserved, known and secret, closed and open, private and public, and masculine and feminine.[4]

The history of the relationships between women of different social classes and their connections to books and domestic and social spaces has attracted a burgeoning amount of attention in recent years. The central message of that scholarship is that the experience of early modern women cannot be captured by focusing on a few exceptional individuals. Few women who were not wealthy and well educated have left significant historical footprints, however, and historians like Sara Mendelson, Patricia Crawford, and Laura Gowing have expended prodigious amounts of time and energy unearthing their lives from the fragments in legal, historical, and religious archives.[5] Paula Findlen, Margaret Pelling, and others have developed this work in hopes of capturing the roles played by women in natural philosophic investigation and medicine, and the place of domestic medical manuals, household recipe books, and oral communication in female medical practice has earned a great deal of attention in recent years.[6]

Written sources, including letters, diaries, and manuscript books, have supplied information about the experiences and expectations of wealthy women with the capacity and desire to write. They do not, however, shed a great deal of light on middling and poor women who were more likely to be able to read and have access to books than to write and have consistent access to writing materials. Inexpensive print, including small books, pamphlets, and ballads, provides one avenue for accessing this group of women, but they require a different kind of interpretation and supply a type of information different from that in manuscript sources. They reflect cultural expectations for women and femininity in the household and in public, cultural anxieties about gender transgression and often the nature of that transgression, the cultural positions held and roles played by women, and the relationships among different classes of women, between women and men, and between women and the natural world.

Changes in the English social and economic structure, particularly shifts in population density and work habits, also contributed to a new emphasis in the public imagination on gender conformity. Appropriate female behavior was the topic of a widespread and longstanding popular debate that spawned an endless number of ballads and pamphlets, some written by women of various social stations and some by men. David Underdown and Laura Gowing, among others, have devoted scholarly attention to the ways in which femininity was constructed and debated in the public forum.[7] This chapter will expand that body of work by attending to constructions of femininity in books of secrets.

Making Secrets Feminine

All three of these books faced two kinds of risk in their potential for reaching an audience of female readers, and each kind of risk was, in its own right, gendered. These books needed to appear both sufficiently useful and innocent to overcome the anxieties associated with telling women secrets. They also had to be packaged in a style familiar to women, so that female readers would feel comfortable with the books and confident enough to purchase and make use of them. One strategy, the presentation of material in straightforward, step-by-step recipes, contributed to both goals. "To bake a Fesant, or Capon in steede of a Fesant. Dresse your Capon lyke a Phesant trussed, perboyled a little, and larded with swete lard: put him into the Coffin, cast theron a little Pepper and Salt: put therto halfe a dish of sweete Butter, let it bake for the space of iii howres, and when it is colde: serve it forth for a Phesant."[8]

Recipes already held an established place in the feminine world of print in the shape of cookery and receipt books. Women compiled their own manuscript collections of recipes for food and herbal medicines, they shared recipes with each other and with men, and they purchased printed recipe books.[9] Recipes were an ideal form for books of secrets because they packaged unfamiliar and exotic materials and ideas in a familiar and accessible form.[10] Under these conditions, a recipe for Damask water that required distillation would share the structure and the language of one for preparing pheasant. "A Powder wherewith to make sweete waters. cap xiviii. Take of the wood of Cipers, or the roots of Galingal, .i. quarterne. Of Calamus aromaticus .i. quarterne. Of Drace or Iris one quarterne. Of Storix Calamit, one quarterne. Of Benjamin, one quarterne. Or ye may take of each of these, one ounce for a proportion, let all be beaten into a powder: and when ye wyll distyll your Roses, fyll your Styll with Rose leaves: and a few Spick flowers, and upon the top of them, strew some of your pouders and so distill them."[11]

The recipe style not only presents the process of distillation in the terms of feminine knowledge, it also presumes that the tools and knowledge required to distill are already part of readers' repertoire. While this may not have been true, particularly for relatively poor readers who lacked the funds to purchase this kind of equipment, the seemingly natural inclusion of distillation within the realm of female knowledge contributes to its assimilation as part of the body of knowledge believed acceptable for female consumption and use. The fact that elite women were more likely to have access to the equipment and ingredients required to complete some of the recipes underlines the fact that they were privileged in their relations to natural knowledge and points to the ways in which knowledge and its practice traveled. Women who lacked access to required apparatus could ask to use a local gentrywoman's, since elite women were generally believed to be responsible for the health of the people in their parish, while elite women who lacked the necessary equipment could take their ingredients to an apothecary and have them compounded to learn the steps and determine whether they actually wanted a still.

Recipes were not only useful for rendering unfamiliar knowledge and ingredients familiar and making them feminine, they were also powerful tools for rendering the natural world predictable and manageable. The recipes in these books present steps to be performed in temporal order, and they introduce a model of the world in which following the instructions with the listed ingredients will always produce the same outcome.[12] They bring the natural world under control in the feminine realm of the kitchen using an interesting combination of techniques and tools gathered from both masculine and feminine bodies of knowledge, particularly alchemy and cookery. These two disciplines already shared the recipe as a form of communicating information about preparing specific substances, and the belief that recipes when performed correctly dependably produce the same results. Alchemical writing had complicated this set of expectations by claiming that recipes for powerful substances like the philosopher's stone could discern between diligent men of pure heart and good intent and dilettantes, because only the former would be able to correctly follow them. Recipes for less lofty substances such as quince tart and capon also privileged diligent readers and practitioners over others. The title page of *The Treasurie of hidden Secrets,* for example, privileges "all women that love and professe the practice of good huswifery, as well wives as Maides."[13] The dedication to good housewifery, like a dedication to good alchemy, guarantees that the information contained in the text will stay secret, or at least opaque, to selfish readers and dilettantes.

Like the books of secrets already discussed, those intended for female readers have the difficult task of promising that the information they contain is novel and useful as well as accessible and affordable. In order to do so, books of secrets make the complete revelation of their contents depend upon readers' intentions and characteristics. Books of secrets for women, however, have a more difficult task because of popular conceptions of the relationship between women and gossip. Women were widely believed to be unworthy of holding privileged knowledge because of a tendency toward mindless chatter. At the same time, their place in birthing rooms and as those responsible for their own and their families' health determined that they were already in possession of secrets that many men would have to dissect to know. As a result of this dual conception of women as both holders and revealers of secrets, women interested in learning about and manipulating the natural world operated in a highly charged realm. They were held back from studying nature as it was accomplished in the universities and were restricted instead to areas of natural knowledge that fell within appropriate boundaries of female responsibilities, particularly herbal medicine and cookery. Within these two realms, however, women were exposed to ideas about change and vulnerability, including systems of sympathy and antipathy that were part of the broader study of the natural world. The presumed relationship between femininity and speech, however, made the secrets printed for female readers part of a broader arena of discourse than those printed for male readers.

These books borrow from and contribute to the long female tradition of oral communication on matters of housewifery, husbandry, cookery, medicine, and gardening.[14] They underline that connection by playing with the word "secrets," making use of its definition as hidden, mysterious, or unusual information about the natural world to make the information they contain worthy of gossip. And because their contents are generally restricted to topics that were suitable for female consumption, they contribute to the project of gendering female readers by teaching them more about the arenas in which they were expected to excel. This aspect of books of secrets intended for female readers takes advantage of the gender expectations surrounding women and speech to make the information women learn and discuss appropriate and useful, rather than potentially destructive. Secrets in this iteration are useful to the feminine projects of making better food and caring for bodies afflicted by diseases and accidents, and their revelation to unworthy readers will either go unappreciated or possibly contribute to their readers' improvement, rather than threatening the body of scholarship that produced them. Like the secrets presented in alchemical recipes, they would work only for worthy, which in this case includes appropriately gendered, readers.

The recipe format fits into the material and theoretical structure of books of secrets, and it conveys knowledge across boundaries of class and gender without drawing attention to them. Structure and content were not always sufficient, though, to buffer these small books against the criticism they could attract from those who preferred secrets left unspoken, or at least protected from the prying eyes of women. Specific defenses adorn the beginnings of each of these texts, and a closer reading of them reveals the variety of tensions plaguing printers and booksellers who produced books of secrets for female readers.

The second page of *The Treasurie of commodious Conceits, and hidden Secrets* boasts an elaborate woodcut featuring a gentleman at his desk, quill in hand, writing on parchment with a reference book open before him and a mostly full bookshelf to his immediate left (Figure 5). He wears clothing appropriate to a sixteenth-century man of some wealth, including a hat, ruffled collar, cape, jacket with decorated cuffs, short pants, stockings, and shoes tied with florets that match those at the hem of the pants. The tone of this woodcut places secrets in the domain of a gentleman scholar, a learned man with the time and resources to devote himself to the study of the world around him. It also places them in the masculine space of the study or the male closet, both rooms that men reserved for private work and study. They retreated to these rooms where they kept a variety of books and private papers in order to be alone with their work. The work they did in their studies or closets could be of a more private nature than the study in which they engaged in more formal, public spaces like universities.

This makes closets the ideal place for the pursuit of secrets, since they protected their inhabitants from the eyes and the judgment of their peers. They were also protected from the scrutiny of other members of the household, who probably were not permitted inside men's closets and certainly did not enter the often elaborately locked doors without an invitation.[15] The woodcut of the scholar places secrets within this assertively private space and within the realm of books, a space both removed from the physical practices required to produce the secrets, like distillation, the growth and collection of herbs, and cooking. Women of all classes would have had better access to the means of producing secrets than reading about them, but their relationship to nature meant that they were not supposed to be capable of organizing what they gained from working with plants and beasts into a coherent system of knowledge. The woodcut scholar is an appropriate host to introduce female readers to the world of natural knowledge. His masculinity and social status firmly situate him within the arena of male learning and protect him from accusations of wantonness and charlatanry.

The woodcut scholar associates the information presented in the book with the masculine tradition of learning, and thus separates it from the feminine arena of sharing with other women their personal experiences with the natural

Figure 5. Image of a gentleman scholar taken from the 1573 edition of *The Treasurie of commodious conceits and hidden secrets*, published by Richard Jones. Courtesy of the Huntington Library Art Collection and Botanical Gardens in San Marino, California where it is held in the Rare Books Collection under call number 59164. Further reproduction is prohibited without permission.

world. His appearance guarantees that the contents of the book will qualify as secret, because they come from an inaccessible realm of thinking about and manipulating nature. His apparent social standing, however, guarantees that the secrets in the book are appropriate for female readers whose femininity could be damaged by too much knowledge, or knowledge of the wrong kind. Finally, he is paired with a clear direction for the book's use that defines the places in which women were expected to practice their knowledge of nature: "Not impertinent for every good Huswife to use in her house, amongst her owne Famelie."[16]

This statement points to the ways in which women could appropriately use secrets about the natural world to improve their abilities as wives and mothers without threatening their femininity. It also points to the critical position of privacy in early modern constructions of femininity, which made it possible for women to know and practice natural knowledge at home but not in the broader world without threat to their gender and reputation. The woodcut scholar polices this boundary. His judgment determines which secrets will be included and which rejected, since it is his work that locates them and his pen that records them. He stands as the compiler referred to on the frontispieces of the 1573 and the 1596 books in the phrase "Gathered out of sundry experiments lately practiced by men of great knowledge." He is not, however, a teller of secrets, a position that would threaten his masculinity and social status by placing him on the same level as talkative, unlearned women. He is a scribe and an "author," relying on the distance provided by his pen and the texts in front of and beside him to legitimize his enterprise and reveal his secrets, placing him at a safe remove from the manual labor of their discovery and the oral aspects of revelation.

While the woodcut gentleman appears only in the first text, gentlewomen figure in the letters introducing all three. In *The Treasurie of commodious Conceits, and hidden Secrets*, John Partridge writes to Richard Wistow, the man to whom the book is dedicated, that the text would not have existed without the prompting of a particular gentlewoman who encouraged him to embrace an alternative system for valuing secrets.[17] He tells Wistow that he collected the secrets in the book with the intention of keeping them for himself and his close friends, "yet at the influence of a certayne Gentlewoman (being my dere and speciall frende) I was constrained to publish the same, and considered with my selfe the saying of the wise: which is, "That good is best, which to all indifferently, is of like goodness, or effect: or which without respect of person, is good to all indifferently."[18]

In this case, Partridge claims that the gentlewoman not only urged him to make his book widely available, she also encouraged him to reconsider the rules behind collecting secrets. He begins his letter by explaining how he began his project, "collecting certayne hidden Secretes together" and compiling them into a single source. By emphasizing the hidden aspects of the knowledge he collected, he reinforces the traditional definition of secrecy and places his book among other such collections. His decision to print, however, depends upon abandoning that definition of secrecy at the behest of a woman on the grounds that secrets are only as valuable as they are useful and accessible.

The feminine influence outlined in this letter works to moderate the masculine desire to collect and horde knowledge for personal benefit, and it disrupts the foundation of the most common method for evaluating secrets. This role for the feminine depends upon the gendered notion of women as secret tellers, and it extends that role from the oral realm of conversation and gossip to the printed page.[19] Print does not, however, necessarily connote silent reading, but instead admits books, rather than single recipes or bits of household advice, into the predominantly female arena of domestic speech.[20] The anonymous gentlewoman acts as the secrets' agent in the world of cheap print, rendering her and her concept of utility responsible for the production of the book and any potential misuse of it.

The Widowes treasure owes its creation and its suggested model for use to another anonymous gentlewoman, who earned her reputation as a healer by using its secrets. "This Pamphlet being written (not many yeeres past as it should seeme) at the earneste requeste and suite of a Gentlewoman in the Countrye for her private use, which by these singular practices hath obtained such fame, that her name shall be remembered forever to the posteritie."[21] Her name remains unstated, a secret of its own, while her rumored reputation and social position as gentlewoman and widow act as guides for the book's printing and use.[22] They also establish her authority as a manipulator of the natural world and practitioner of secrets, since the knowledge she gained in the pursuit of her career has demonstrated utility.

The letter gives the manuscript book a life of its own, marking its path from the widow to a friend to the current author. "The original Copie (by great chance) was lent me by an especiall freende of mine, in the perusing whereof I found it so furnished with rare experiments and pretty conceits as the like are not to my knowledge extant in the English tongue."[23] The pretty conceits that so impressed the letter's author were even less secret to most women than were the connections between plants and animals described in *Albertus*

Magnus, although they would not have occupied significant space in printed books, academic culture, or men's brains. They did, however, certainly fill a multitude of manuscript receipt books and women's heads. These secrets formed the bedrock of the primarily female world centered on satisfying the private demands of domestic life.

The letter's author acknowledges the private, female origin of this book of secrets by telling the story of its creation at the widow's behalf, but he skirts the shared domestic demands that spawned countless private collections of the same kind among a broad range of women. Since his project for the book focused on its inexpensive production and public distribution, acknowledging the possibility of countless similar manuscripts would have been counterproductive. He sets about establishing this book as unique, inherently different from private knowledge collections because of its printed format and professional roots, and perfectly suited to public use. He accomplishes this task by pointing to the widow's reputation as a skilled healer: "This I dare presume to reporte of the woorke, that there are herein included very manye secrets, that I knowe by the Widowes owne practise, to be most singuler and approved."[24]

The act of witnessing testifies to the originality and functionality of the secrets included in the book. It also establishes the widow as a public practitioner of medicine and natural manipulation whose expertise can be witnessed. Her status as a nonpublic character is not, however, sufficient to warrant a first person appearance, and in the same way that the author cites her successful practice as evidence to support the book's entrée into the public print market, he cites her success to support her presence in the professional marketplace of secrets.

The widow's appearances in the letter work together to establish the book as private and unique, to attest to its utility in public medical practice, and finally to move it from a private project founded in personal desire to a publicly available commodity desirable to other women. The letter links the book's move from private to public use and space to the widow's move from being an expert in medicine and cookery at home to practicing the same arts in public. The widow's final role in the introduction of this text is spoken. She makes a claim for her book's proper use through the letter's author, in which she defines "good readers" as careful and diligent, just like the "good readers" described in other books of secrets. "This caveat she also giveth by the way, to read them advisedly and practice with discretion."[25] Like *Albertus Magnus, Cornucopiae,* and *The Mirror of Alchimy,* this book defines a good reader and practitioner of secrets through the assignment of specific characteristics. In this case, an awareness of the fraught nature of women manipulating nature, even if that manipulation produces jam instead of the philosopher's stone, heightens the importance of prudence in readers' personalities.

The widow has significant power in the introductory letter because her reputation as a successful practitioner of secrets and gentlewoman justifies the book's traversing of the private-public divide. The image of the widow, taken in the context of the letter, suggests that anonymous women of secure social position and established accomplishment as manipulators of nature make good partners for male authors and printers who wish to print books that tell women how to use secret knowledge to better inhabit and improve their private spaces. Neither the male compiler and printer nor the female source of these secrets is strong enough to withstand public criticism, and the combined weight of their intellectual and social standing is brought to bear to enlist readers in defending and adopting the project. The author's voice reappears in the last lines of the letter, this time speaking for the widow to ask that readers protect them both as the book passes from the private circle of conversation and reputation into the public arena of gossip and fame: "I dedicate unto you this her Treasure and chief Jewell: desiring you to be a defense as well for her as for me, from all suche as shall in your hearing scorne her or mee for disclosing such profitable experiments."[26]

The themes of establishing authority and guidelines for proper use and justifying the presence of female practitioners using the secrets of nature in public continue in *The Treasurie of hidden Secrets*. It places even greater emphasis on defining the characteristics that make someone a good reader and user of secrets. The introductory letter, written by the printer, is highly specific about the best places and ways to use the book. Richard Jones asserts that the secrets he presents are "Not impertinent for every good Huswife to use in her house, amongst her own Famelie."[27] Readers were, of course, free to ignore Jones's admonitions, and some probably did. His anxiety over the matter seems disproportionate to the limited potential for gender anarchy posed by recipes for quince tart, particularly in light of the fact that women were generally responsible for the health of everyone in their household and, in the case of noblewomen, the poor of the town. Fulfilling their social roles necessitated that women of the upper classes dispense medical advice and remedies outside their homes, and it seems improbable that Jones wrote his disclaimer solely in the name of limiting women's medical work to their homes. Instead, his letter defines acceptable work for women by reinscribing a divide between the private and the public, where women, cookery, household duties, and domestic medicine are part of the private sphere, even when that sphere has to expand to include public duties.

By packaging and selling a book of secrets for women, Jones undermines the privacy of this kind of knowledge. He also erodes its status as secret by making it available to a broad audience for an inexpensive price without a guarantee that this knowledge would remain in women's heads and homes. Like the secrets in the books discussed in chapters 2 and 3, the information he conveys is not necessarily unknown to readers. Unlike the secrets in those books, which offer broad control of the natural world through the manipulation or imitation of natural relationships, these rewrite the natural world in the form of recipes and, in so doing, make it predictable.

The secrets in this book play the same role as D. A. Miller's "open secrets," becoming information known and exchanged under the heading of secrecy that acts not to protect its contents, but to mark, and sometimes privilege, them.[28] In this case, the recipes in the book are defined as secrets in order to guarantee them a place among other exchanges of natural knowledge. They are also marked as feminine, indicating that they derive from the private domain of the kitchen, the garden, and the bedside, where recipes govern the health of every member of the household by determining diet, a significant aspect of humoral medicine, and governing the preparation and administration of medicines. They are privileged knowledge because they traverse the divide between the public and the private and invite readers to learn domestic knowledge and, in the process, bring into control their and their families' bodies.

Richard Jones's effort to delineate the proper use of this book gives some evidence of how the book could operate in readers' hands. The letter, note, and poem that introduce this volume complicate the problem by addressing its highly convoluted authorship and establishing the knowledge space it is intended to occupy. The first few sentences of Richard Jones's letter to his readers claim the book as his own creation, since it represents the combination and reorganization of two books he had previously printed. "I have therefore placed each thing that before was out of order in his due and convenient place, and doe commende both unto your protection: the one for your kitchin, and this other a readie help, always at hand as a Storehouse, or Treasurie of manie profitable secretes, and unknown Conceites to be used as occasion shall require."[29] These lines emphasize the preeminent position of organization in determining authorship of secrets, since disorganized secrets would be destined to remain unappreciated by the majority of readers.

The next sentence, however, reminds readers of the anonymous gentlewoman who requested the printing of the first collection of secrets and who remains crucial in justifying the revelation of secrets by tying their value to their general utility. The figure of the gentlewoman dominates the conversation between the author and his book, a short note that is a new addition since the 1573 printing. Her appearance destabilizes Richard Jones's claim to the position of sole organizer, and thus author, since it introduces another compiler and draws attention to the "Printer" that follows Jones's initials at the end of the letter.

The gentlewoman's friend was probably John Partridge, since the earlier book was attributed to him and makes up a good part of the new version. In this edition, his connection to the book is again determined by the lady's request, but the request has changed. Rather than asking that a private collection of secrets be made public, she has asked for her own copy of a book of secrets. "Upon occasion that a Ladie of honorable regard, having seene this booke in writing, earnestlie requested, or rather commanded, to have a copie of the same."[30] This shift is important, since it makes the lady responsible only for the creation of a single copy, the first whispering of the secrets, and Jones responsible for the widespread distribution of the book.

The poem that ends the series of introductory pieces cements the lady's role from the earlier book as the moving force behind the decision to print and the one who caused the secrets to be told. It concludes with the couplet "Yea, say her command of me hath obtained/Thee: that no golde nor good could have gained."[31] She becomes responsible for the transition from secret to open secret, and her role as a transgressor of the divide between the public and the private is underlined by her very feminine desire for knowledge. The knowledge she causes to be revealed, however, reinforces the specific ways in which women, from their domestic locations in kitchens and gardens, can manipulate the natural world.

The Secrets and the Text

A tremendous amount of effort and thought was devoted to packaging and presenting these small books as appealing, entertaining, and useful additions to women's collections. Each book bears some important similarities to the others, with the first and last to be published having more in common with each other than either has with *The Widowes treasure*. This is not particularly surprising, since *The Treasurie of hidden Secrets* simply combines the earlier *Treasurie* with another domestic manual. The depictions of gendered work, secrets, and the natural world also provide some insight into the ways in which these books could be used by women to build and maintain healthier, cleaner, and more predictable meals, bodies, and households, which could, in turn, support their belief in an ordered and orderly universe.

The Treasurie of commodious Conceits, and hidden Secrets, printed in 1573, presents 68 recipes in 95 pages. It begins with the scholar woodcut opposite the frontispiece and devotes 7 pages to introductory remarks about the text and letters supporting it. The table of contents lists all the entries in order and comprises 5 pages. The recipes begin after a final poem from the author to his book, and the structure of the text remains relatively consistent from that point, with variations generally in the form of additions to the recipe rather than changes in its presentation. Each recipe begins with a heading and chapter number, which is followed by the list of ingredients and instructions for preparation. The seventeenth recipe, which gives directions for making conserve of roses, is the first to present any change to this basic structure.

The recipe is followed by a separate paragraph on the conserve's virtues, which is included as part of the original chapter number. After learning how to prepare conserve of roses, readers also learn that "Conserve of Roses comforteth the stomack, the heart and all the bowels, it mosyseth and softneth the bowels, and is good againste blacke Coler: melancholy, conserves of white roses doth loose the belly more than red."[32] This recipe and list of virtues is followed by a similar one with instructions for making conserve of violets with variations in the amount of sugar to be added when making conserves from different kinds of flowers. The next eight entries actually present only the virtues of various floral conserves, with the reader expected to refer back to the original recipe for conserve of violets to make any of the other versions.[33] This section of the text ends with a summary of the approaches required to make conserves of roots, fruits, and flowers and a reminder that adding sugar, as long as it remains in the correct ratio with the other ingredients, will only improve the results.

Eleven recipes later, the structure of the text changes again. Rather than a simple heading, chapters 37 through 39 advertise what the product is intended to accomplish along with its name. Thus, chapter 37 is for "Powder of Holland against Colick" while chapter 38 is for a nameless "Powder to make the belly soluble" and chapter 39 "A receipt to restore strength in them that arr brought low with long sicknesse."[34] The rest of the text alternates between headings that offer only the name of what is to be made and those that list its uses.

The first entry to deviate from the recipe model comes in the final quarter of the text and signals a transition away from a mixture of culinary and medical recipes to a strictly medical focus. It presents guidelines for the collection of herbs, seeds, and flowers to ensure their full strength for medicinal uses, with an emphasis on location and timing in the plant's life cycle. The importance of this information is underlined by the warning "Many Herbes there be that have special time to be gathered in: And if they bee gathered in their time, [the]y have their whole vertue to their propertie, or els not so good."[35] The final piece of this section is devoted to a list of important medicinal plants and the times of the year that they should be gathered, with a brief entry for each listed plant.

The next section of the text begins with "the sundrie virtues of Roses for dyvers Medicines" and offers the virtues of lilies, milfoil [yarrow], and rosemary as well. Each of these entries reiterates some of the merits mentioned in the earlier section dedicated to making conserves, but they include a great deal more detail, with the virtues of rosemary extending some 4½ pages and covering its benefits for both sexes and a broad range of medical problems.

"A briefe Treatise of Urines" comes next, and it begins with a paragraph diagramming the anatomical location of sickness and health. "That is: in the wombe, in the head, in the liver, and in the blather: in what maner thou maist know their properties, and thereof thou mayst learne."[36] The products of the bladder are chronicled in detail for the next three and a half pages, covering a range of urines of different colors and textures denoting different diseases in both men and women.

The final entry in the text is for Dr. Stevens's water, a sovereign remedy guaranteed to cure barrenness in women, palsy, worms, toothache, gout, dropsy, upset stomach, bladder stones, reins, canker, and bad breath. Its efficacy not only guaranteed Dr. Stevens's successful career and many miraculous cures but also ensured his long, if rather sedentary, life: "It preserved Dr Stevens that he lived ixxx and xviii yeares whereof x he lived bedred."[37]

The second *Treasurie,* printed in 1596, is similar to the first, but it is more strictly organized and more evidently a part of the cheap print genre. Gone is the elaborate woodcut of the first edition, the letters from distinguished men testifying to the author's character, and the lengthy dedication. The type in the second edition is much smaller, and the spaces between recipes are minimized in order to squeeze as much information as possible into fewer pages. All these space-saving measures result in a book that is 23 pages shorter than the first but contains 149 recipes, rather than 68. The final structural difference is the table of contents, which appears at the end, rather than the beginning, of this text.

The first difference in content between the two texts occurs in the culinary recipes, with the second *Treasurie* having no recipes for meat dishes, which comprised a significant part of the cookery recipes in the first text. It features instead lengthy instructions for preparing conserves and baked goods. The second *Treasurie* also contains a number of new recipes, the majority of which address medical problems, with a focus on women's health. It features a new section subtitled "secret remedies appertaining to women," which includes recipes for bringing on menstruation, limiting menstrual bleeding, treating the falling sickness and suffocation of the matrix, delivering a dead child, delivering the placenta, and treating fainting in childbirth.[38] Scattered throughout the book are more recipes on women's health, including how to determine whether a woman is barren, how to make barren women bear children, how to speed labor and delivery, how to treat breasts sore from milk, and how to cure a woman's gnawing stomach.

The second significant addition comes from another tradition of secrets, which tries to anticipate the bad things that can happen, especially in strange places, and provide remedies for them. The recipes that fall under this heading include one intended to drive all venomous beasts from your home, one for drawing arrowheads or other iron out of a wound, and one for curing all poison eaten or drunk.[39] The final addition fits into the communication of academic or systematized knowledge explored in the first text through the treatise on urines. It includes instructions on when and where to gather herbs to maximize their power, the names and natures of different diseases, directions for distilling and advice for what equipment works best, and instructions to guide reactions to various physical and astronomical observations.[40]

All these subsections enlarge the theoretical basis of the text and the recipes included in it, expanding the knowledge covered by its second version. It remains, however, a book dedicated to the education of housewives and the improvement of private spaces, and its ordered approach to the task creates an image of an orderly world that lends itself to structuring strategies like tables of contents, subsections on various topics, and brief theoretical treatises. Like the structuring strategies in *The Secrets of Albertus Magnus*, these approaches to categorizing natural knowledge place the readers' desires at the center of the text, making it useful as a reference tool, and allow it to be read as a whole, which produces an image of the natural world that is predictable and accessible to readers. The image of the world produced by *The Widowes treasure* is quite different, and the strategies used to tame it are smaller in scale, at the level of the recipe rather than at the level of the text. Like *Cornucopiae*, it resists narrative progression in favor of the repetition of categories and relationships that together compose the natural world. Unlike *Cornucopiae*, however, it uses general categories to structure its image of nature rather than depend upon relationships between entries to determine the world's composition.

The Widowes treasure is 113 pages long and contains 266 entries. The vast majority of these fall into three categories: cookery, human medicine, and veterinary medicine. The remainder could be considered nice things to know, such as how to make inks of various colors and how to test whether the wine has been watered. There is no table of contents, and so readers would have had to mark or annotate their editions, copy down useful recipes into their recipe books, or have extraordinary memories to make use of the text. This suggests that it would have made a poor reference tool and was intended, like *Cornucopiae*, for instruction and amusement.

The text looks more familiar than it is, in that its structure mirrors the headed entries that fill the two *Treasuries*. It begins with the familiar offering of a recipe for syrup of roses or violets, followed by one for diatreon piperion (an aid to digestion) and a poem touting the virtues and listing directions for the correct use of this remedy. It begins, "This decoction is good to eate,/always before and after meate./For it will make digestion good,/and turne your meate to puer bloode."[41] The poem comes in pairs of easily memorized rhyming couplets and makes use of a different kind of reading strategy than the one needed to follow a recipe. Catechism was the method by which many early modern readers became literate, and the rhyming couplets in this poem would have made use of readers' trained memories to teach them when and why to use this particular remedy. The poem ironically references the importance of memory, making use of the remedies' qualities to remind readers how to use both the medicine and the poem to help themselves: "Besides all this it will restore,/your memory though lost before."[42]

The poem is an aberration in this text, not only because of its unusual form but because it explains the correct usage of the remedy created by the preceding recipe. The remaining recipes either give no indication of how and when they should be used or advertise their uses in their headings. The latter is by far the more common strategy, and readers would have experienced a vast list of remedies, sovereign and generic, headed by the medical or veterinary problems they were intended to cure. The text is not entirely without internal structure, as the recipes are loosely grouped together by the species requiring medical attention—that is, all the veterinary recipes appear together, with the first set of recipes for horses and the second for cattle. The cookery recipes tend to fall together as well, with lengthy entries devoted to the making of jellies and conserves and several pages on the proper preparation of various meats, vegetables, and eggs.[43]

The medical recipes also appear in clumps, though the organizational principle behind their groupings is sometimes difficult to detect—for example, it seems sensible for the two recipes treating flume in the stomach to appear back to back, as other recipes for the same problem do, but they are separated by two recipes giving instructions for preparing remedies for unrelated ailments.[44] Rather than detracting from the natural order produced by the text, however, lapses like these suggest a highly varied, complicated, textured world that bursts through attempts at strict ordering and insists instead upon disorder. The march of recipes in the face of this physical disorder suggests a means of coping with it and keeping it under control at the individual, bodily, and household level that makes the imposition and maintenance of order a feminine responsibility.

The nature of the recipes swerves and turns on itself, covering a broad range of physical ailments suffered by men and women, interspersed with directions for preparing quince marmalade, dressing a hare in broth with pudding in its belly, and helping horses grow strong hooves. This range represents the daily experiences of early modern women whose lives ranged from the kitchen to the garden to the stable yard, and could even, when they had a spare moment, include the enjoyable task of preparing and using inks of different colors. The unruliness of this text captures the unruliness of the worlds that women inhabited, while its steady flow of answers to the questions posed by those worlds suggests that recipes had more to offer readers than an answer to a specific problem. This book and the recipes it contained constituted a strategy for managing and ordering the seemingly unmanageable and disordered natural world, while the two *Treasuries* presented orderly worlds in which recipes helped to address imbalance created by readers' ignorance about or confusion over the ways to best meet their families' needs.

Gendering Natural Knowledge and Producing Natural Worlds

The world created by the first *Treasurie* can best be imagined as existing within a housewife's kitchen. Most of the recipes and instructions contained in this text refer tangentially to the place in which they will be prepared by listing the tools, ingredients, and activities required to make them. "To bake

Quinces," for example, "Pare them, take out the core, perboyle them in water tyll they be tender, let the water run from them til they be drie."[45] Baking, paring, parbroiling, and draining are all activities performed in kitchens, and while the recipe never names the space, it is evident from the word choices and the subject matter where preparation should take place.

This is not just true of the culinary recipes, which have obvious links to the kitchen. Even the medical recipes seem more tightly linked to the process and place of creation than the proper moment or method to use them. "To make Conserve of Roses, Take the buddes of Red Roses, somewhat before they be ready to spred: cut the red part of the leaves from the white, then take the red leaves and beat and grind them in a stone morter with a pestell of wood, and to every ounce of Roses, put iii ounces of suger in the grinding (after the leves ar wel beaten) and grinde them together till they be perfectlye incorporated."[46] The verbs and the tools mentioned in this recipe place its creation and the reader within the feminine space of the kitchen and engage in controlling disorder in the natural world from that space. The instructions for preparing this recipe are more than twice as long as the list of its virtues, and there is no indication of how the remedy should be applied, just the ailments it effectively treats.

The majority of the medical recipes look like this one, a set of instructions followed by the ailments the recipe can remedy. There are, of course, a few exceptions, which are noticeable because they are always headed by their uses, "To make a fumigation for a presse of clothes" or "For the ague in a woman's brest," rather than "To make Conserve of Gladon, or yellow Flowerdelice: With the sundrie Vertues and Operations thereof."[47] While these headings emphasize the remedies' uses over the processes required to make them, the act of preparation remains critical, as indicated by the frequent appearance of the verb "to make." Women's acts of creation occurred in the kitchen, where they kept the tools and ingredients required to make a range of household products, and entries introduced by an impetus to create tied readers and the text to the space where that act could most easily occur.

The biggest departures from this model require that readers leave the mostly private and feminine space of the kitchen for less well defined spaces. These entries are advertised by the new verbs used in their headings: "To *know* what time in the yeare Herbes and Flowres should be gathered in their full strength" and "A briefe Treatise of Urines as well of mennes urines, as of womens, to *judge* by the colors, which betoken helth, which betoken sickenesse, and which also betoken death [emphasis mine]."[48] These verbs, which insist upon intellectual rather than physical activity, move women from the action-defined space of the kitchen to larger, more public spaces like fields or theory laden, private spaces like sickrooms. These two entries, which come at the end of the text, acknowledge the potential for female readers to reach beyond the practical spaces of the kitchen and the sickroom to the intellectual spaces of herbal and medical theory.

The order of the entries is important, since the directions on when and where to collect medicinal plants informs the herbal content permeating the rest of the text. It offers readers a short step away from a textual world in which individual plants exist as ingredients in specific recipes to one in which plants coexist with each other in a system designed to permit readers to maximize their utility across recipes. Roses, for example, are mentioned in various forms as ingredients in eleven recipes before instructions are given for their proper collection and a full list is made of their medicinal virtues.[49] This entry is tightly tied to the rest of the text, and while it beckons readers away from the timeless confines of their kitchens to the fields and seasons in which herbs are collected, it does so to improve readers' ability to perform feminine duties in feminine spaces.

Roses, despite all their previous appearances, had never been characterized as "colde and moyste, in two degrees."[50] By integrating them into the theoretical framework of humoral theory, roses become a bridge for female readers between what they already know about specific diseases and a general system for thinking about all diseases and therapies. The treatise on urines builds on that theoretical infrastructure to empower readers' judgment. Readers, after all, were the ones expected to observe urine specimens, match them to the descriptions in the text, and work from the predictions made in this section to determine the correct course of treatment. The most remarkable thing about the treatise on urines is that it says absolutely nothing about the kind of treatment suggested by each condition, implying that readers were expected to work backward to determine which recipes, or even individual herbs, would work best to treat each problem. It even invites readers who identified a problem for which no recipe existed in the text to build their own, using the information provided in the section on the proper collection of different plants and their virtues. The creative act required here is intellectual, not manual, and makes use of the kitchen only to satisfy the practical demands of making a remedy. It provides readers with the potential to imagine and control the natural world through creative interpretation rather than rote memorization, and it moves them beyond the walls of their kitchens into the intellectual space required to inform their judgment and allow them to make the best use of available tools and ingredients.

The Widowes treasure also begins in the kitchen and it remains in spaces commonly accepted as feminine, such as the sickroom, the garden, and the barnyard. The challenge posed by this text is not the physical or intellectual borders that it encourages readers to transgress, but rather the level of caretaking and highly varied responsibilities that it assumes for its readers. These responsibilities match the categories covered by the recipes: food preparation, human bodies, and animal bodies. Each of these categories has a proliferation of things that could go awry, and the recipes aim to restore order where it has been lost and add pleasant touches to daily living. The culinary recipes and those for the preparation of various colors of ink both serve this purpose, with the latter more evidently devoted to pleasure and amusement.[51]

The recipes for food in this text, for example, pay attention to making food enjoyable as well as palatable and nourishing. The recipe for mutton is an excellent example, in that it moves beyond correct preparation of the dish to reminders for proper seasoning and serving. Once the mutton is boiling with some toasted bread soaked in wine and vinegar, "Put in some Pepper, whole Cloves, and Sinamon, and let all this be well stewed, but forget not Salte. Serve it out upon soppes: and boyle some Onions in it cut into pieces."[52] Whatever your opinion of boiled mutton, the attention paid to make this dish more pleasant indicates that food preparation was a place where women could be creative to nurture their households beyond the basic satisfaction of bodily needs.

Daily life is certainly one of the orders monitored by this text. In the recipe "To provoke sleep," ingredients and routine are marshaled against the disruption of insomnia. "Take a spoonefull of Womans milke, a spoonefull of Rosewater, and a Spoonefull of the juice of Lettice, boyle them in a dishe, then take some fine Flaxe, and make your plaister as broade as you will have it lye on your forehead."[53] The ingredients in this recipe draw entirely from female spaces, including maternal bodies. This neat circular reference, in which the maternal figure of the housewife returns to the sustenance of her childhood and the products of her kitchen garden and of her loom to regain bodily rhythm reinforces the textual message of women as caretakers and restorers of order.

Of course, female bodies were not the only ones for which readers were responsible, and the personal note of the insomnia recipe falls away when a recipe is given for a specifically male problem. The recipe "For the swelling of the Yearde or Coddes" urges readers to prepare a solution to bathe the affected body parts, and the verbs it employs to discuss that process are remarkable for their caring qualities: "then with a ragge of Linnen cloath annoynt the Yeard under the skinne with the water well and warme: and it wil abate the payne of the Yearde and of the Cods also if they be washed with the same."[54] Readers would likely have been treating afflicted family members toward whom they already felt a sense of care and responsibility, but the tenderness encouraged by the wording of this recipe serves as a reminder of the role that women played in nurturing others' bodies and protecting their dignity in the process.

The nurturing tone of this recipe is also present in some of the recipes to combat veterinary ailments. The list of common problems for horses and how to treat them with household ingredients runs for sixteen entries and includes recipes to treat worms, bots, hooves of poor quality, saddle galls, sore backs, diarrhea, horse leeches, punctures from nails, and snake bites. The success of these remedies would have depended upon a close and trusting relationship between reader and horse, a relationship that is supported by the pieces of advice sprinkled throughout the recipes that recommend care beyond the minimum required to get the job done.

The recipe for "A drench to plume up a horse, and to expel colde, to clear him of the Glaunders, and to open the pipes" is an excellent example of the ethic of caring embedded in the narrative. The instructions to make the drench are followed by advice on its administration: "Give him this drinke luke warme a quart at a time, then ride him till he be hot, then bring him into the Stable, and litter him well, and corie him till he be thorow colde, then let him after stand upon the bitte, and turne him to grasse."[55] The directions for caring for the horse after riding can be considered a crash course in equine coddling, since they encourage the reader to carefully cool out her horse while currying (brushing) his coat in a deeply bedded stall, then return him to his field. Two of these steps would have sufficed to prevent him from colicking; the brushing and the fresh stall are simply kind things to do for a hardworking and valuable animal. Here nurturing the animals and people in their care is made a feminine virtue. In the next chapter, we will see how a very different population, gentlemen and their stable boys, provided the same services in the name of producing and maintaining performance and racehorses who served as public evidence of their masculinity. This text places humans and animals in a very different light, with women looking after horses with the same attention that they gave to other beings in their care and spending time and energy watching and tending to their livestock with an eye to keeping them well and happy.

The Treasurie of hidden Secrets builds upon the practical, kitchen-centered world of the first text and the caretaking impulse of the second to combine extensive theoretical treatises with the more typical recipes for food and medicine. It centers itself on the female body, which would come to be the focus of many books of secrets in the seventeenth and eighteenth centuries. In this text, that space was already characterized as secret, and the subheading "Certaine secrete remedies appertaining to women" sets it apart as an enclosed realm of secrets within the larger text. These secrets are different from those contained in the rest of the book, not just because they address a specific kind of body, but because they deal entirely with a part of natural world gone awry, driven to act against its own survival.

Readers of these recipes would have come face to face with wombs that threatened to suffocate and to fall, periods that wouldn't come and periods that wouldn't stop, lengthy and difficult labors, and children born dead. The problems themselves were presented in short, descriptive phrases: "For the suffigation of the Matrice, and for the falling of the same," "For the Flowers to be brought out shortly," which helped to normalize them and bring them within the bounds of the rest of the text, making them seem as manageable a goal as "A sweet powder for Naperie and all Linnen clothes."[56] The clear, straightforward recipes under the headings bring imbalanced bodies under control by placing them in relation to clear instructions that advise observation, the use of known ingredients, and the employment of easy techniques.

The tone of the text never falters from one of steady, stepwise advice, acting as a guide to women as they confront their own or their daughters'/sisters'/friends' bodies. This book could serve to reassure nervous women as they attempted to deal with their first crisis, or it could act as a reminder to experienced practitioners. It could also offer new recipes for addressing well-known problems, an aspect that is illustrated by the tendency to list at least two recipes for most problems. Finally, it underlines the responsibility and power to be good practitioners that accompany the knowledge of women's secrets. The recipe to restore menstruation begins, "First looke that the woman bee not with childe" before they begin to use a recipe that could end a pregnancy.[57] The section on urines supplies a means for discovering whether a woman is pregnant: "Womans Urine that is cleare and shining in the Urinall as siluer, if she cast oft, and if she haue no talent to meat, it betokeneth she is with child."[58]

The secrets in the majority of this text exist without any theoretical support, but three sections lend depth to the intellectual project undertaken here. The groundwork for this level of theoretical or background discussion was laid in the first *Treasurie* in the sections on urines and when and where to pick medicinal herbs to maximize their virtue. Both of these were repeated in the second book, with some information added to the section on herbs. Three new sections contribute a richer, multidimensional intellectual infrastructure to the utilitarian recipes that fill most of the text and expand the knowledge available to and appropriate for female readers. The first section addresses the art of distilling, and it gives instructions for the equipment best suited to the task, how to distill various herbs and flowers, and the order of operations for an effective distillation.

While the first *Treasurie* presents distillation as an accepted part of readers' knowledge, the text ensures that readers will have the proper tools and skills required to perform the recipes requiring distillation, as well as the knowledge to properly handle the products to be distilled. In this piece entitled "Conclusions and rules to be used in distillings and the ordering of each hearbe or flower before they be distilled," the text frames the process of distillation in the same matter-of-fact manner in which it presented the making of jams and the treatment of women's diseases.[59] It serves to place the process, one that had long been part of masculine knowledge, neatly within the arena of feminine knowledge by presenting it through terms associated with the kitchen, like cleaning, boiling, weighing, and measuring, and by addressing it in the context of familiar ingredients like violets and chamomile. This strategy allies the domestic projects of women with the process of distillation, so that the latter seems a necessary and logical part of the former. The transition from arcane technique to mandatory domestic activity takes place in enumerated steps. "First, a soft fire maketh sweet water, and the sweetnes to continue strong. Secondly, coales still the best water. Thirdly, wash nothing that you will still, but wipe it with a clean cloath. Fourthly, all hearbes, flowers, and seeds must be gathered when the dewe is off them."[60] The spoken language of the first four steps, in which you can almost hear an experienced distiller listing the basics of the art, breaks down into numbered commands for steps five through eighteen, then into a list of herbs and the seasons in which they should be distilled.

This degeneration marks a textual transition, or a presumed understanding between the text and the reader, in which distillation can exit the comforting realm of the spoken by depending on the model of the recipe and numbered steps to enter the realm of women's knowledge. The final list of herbs and directions for collecting them can almost be considered afterthoughts to the intellectual project, since they assume that the reader is already prepared to undertake distillation. This list also limits readers' potential to use their stills to attempt alchemical projects, since relatively specific directions are given to support the distillation of herbs and flowers, suggesting that each object requires distinct preparations for the process to be successful, and no directions are given for any metallic products. Again, the text introduces a project and limits it to render it appropriate for women by sticking with a simple, recognizable, step-based presentation. In this case, however, that presentation includes a process that had little history in the world of feminine knowledge.

The second piece of the text—like the first, devoted to filling in the background of readers' knowledge—comes in the long section disclosing the names and natures of diseases that commonly affect men and women. This section is particularly remarkable for its frequent use of Latin rather than vernacular names for some of the diseases it discusses and for the wide range of information it presents. In a book with five recipes devoted solely to improving skin conditions, it is not surprising that the first long entry in the section on diseases focuses on the family of conditions causing blotches, boils, and blemishes on the skin (both the visible skin and that of the stomach and lungs). "Postema is in divers maners both within the skin and without the skinne and within the body, for all maner of things that swell beare out the flesh, and therefore all maner of Boyles, Botches, Fellons, and other such like may be called postema as well as those that be upon the stomack, or on the lungs."[61] This type of problem holds a significant place in the text, and presumably in readers' experiences, and so the lengthy discussion of postema serves to fill in their observational knowledge with information about the condition itself.

The lengthiest entries address another of the principal concerns of the text, female health—in particular, menstruation and childbirth. The entry on menstruation includes the only anatomical description of the entire text. "But the Matrice of it selfe is like a three cornered purse, as it may be made in figure: And that hangeth by certaine strings by the ribs, and by the intrailes, and so it stretcheth downe to the priuy member, which is called the mouth of the Matrice."[62] The anatomical detail is justified by the section that follows, which includes a description of various problems that can occur during birth that, for correct treatment, require the practitioner to be able to distinguish between the womb and the afterbirth.

That section names Gilbertus as its source and sets up his academic and textual authority opposite the practical, applied knowledge of skilled midwives who have to cope with the multitude of problems that can accompany birth. This comparison, in which established authority is invoked to provide the theoretical causes of delayed delivery and practical authority interprets those causes and attends to the resulting problems, demonstrates the liminal space occupied by this section of the text. In a book largely devoted to the practical explication of recipes, a theoretical section must be immediately placed in the context of its practical results.

The section on astronomical observations is also placed within the context of practical medical care, and it begins with a basic principle of astrology, a brief description of the motions of the moon. "Moreouer it is to be understood, that euery moneth in the year the Moone hath her course in one of the twelue signes: and in euery signe the Moone is two days and a halfe almost."[63] This description could be drawn from any elementary astrological text, but it has to be available to readers in order to allow them to make sense of and use the astrological information that follows. "And ye shall know also that the twelue signes haue gouernment of euery man and beast in the twelue parts of the body. And while the Moone is in euery signe: and if the body be let blood or els wounded or burnt, all the medicines that are cannot serue in that signe that hath the gouernment of the place of the body, and it is maruell but that the body be soone dead or els distraught for euer."[64] Practitioners unfamiliar with the effects of the stars on bodies could pick up basic knowledge from the introduction to this section, and they could learn specifics from the list of signs that follows, with the body parts controlled by each house.

This model may seem similar to the one used to communicate the process of distillation, since it goes from the general to the specific, but it lacks the careful transition that permitted the entrance of that process into domestic knowledge. Instead, the straightforward language in which this section is written departs from the stepwise order of recipes and moves freely from prose paragraphs to lists, then back to paragraphs. It is most similar to the section on the nature of diseases, which also aims to provide information that will enhance readers' abilities to use the recipes they find in the text. The prose-based approach of these two segments unites them, as neatly as does their proximity, with the section on diseases immediately preceding the one on physical and astronomical observations.

The astrological section departs in one regard: The final paragraph of the section addresses the nature of the calendar, which is central to the discussion of astrology, rather than continuing to focus on smaller and more specialized pieces of this area of investigation. It begins with a list of the dog days, on which it is dangerous to let blood, and then continues by discussing the naming of days and how to determine when a day has started and when it has passed. "It is to bee understood in the Sunne rising of euery day is the first hour of the Planets: as to accompt after the Planet, the which Planets the dayes were marked in olde time, for in Latin and French they haue taken their names kindly after the Planets, but in English not so."[65]

This piece of the text attempts to reconcile the concept of dog days, which were marked dangerous for medical procedures in antiquity and maintained their power straight through the early modern period, with the English reformed calendar. It does not, however, suggest that the most recent knowledge should replace the wisdom of the ancients. Instead, it praises language systems that maintain a relationship with the Roman naming of the days, scolds the English abdication, and translates the new calendar as closely as possible onto the old. "And in this order ye should accompt the houres of euery day by the Planets. If it be Saturday, the houre at Sun rising, which that day gouerneth, by Saturnus, the next day after by Iupiter. And so in order recken out the seuen Planets. And rise as oft as needeth until the foure and twenty howres bee fully spent. And this is the course of all the Planets of all the yeere."[66]

It is particularly interesting that this negotiation between ancient and new knowledge systems takes place in the context of a book of secrets written for women. Acknowledgments of the breaks within knowledge systems, particularly in medicine, were still very rare, and for them to be both acknowledged and examined in a popular text suggests an extant level of concern about the best way to translate astrological information from the ancient calendar into contemporary medical practice. The presumption that women would be familiar with and actively participating in this negotiation contributes to their presence in this text as practitioners engaged with the problems posed by the natural world and interested in different means for understanding and controlling them.

The final section of the text that deserves some attention addresses an entirely different kind of problem posed by the natural world. These problems stem from a hostile external environment, rather than internal turmoil. Three recipes fall under this heading, one that drives all venomous beasts from your home, one that is a remedy for all poisons consumed, and one that allows the safe withdrawal of an arrowhead or other iron object from a wound.[67] The first two recipes follow each other, suggesting that they fell together at least in Richard Jones's mind when he was constructing the book. The third follows five recipes later, and it may originally have been intended to be read alongside the medical recipes that precede and follow it. It fits, however, with the overall depiction of the external world as hostile and unpredictable and fits with the use of the book to provide remedies for some of the hostile acts that could be produced by that world.

The first remedy imagines animals as the aggressors intent on disturbing a peaceful household. The recipe includes easily found herbs mixed with the suet of a buck and mixed into a paste to be burned within a house when you wish to use it undisturbed. "And when you will use or occupie it, burne it, for whereas the smoke thereof goeth, the beastes will voyd away."[68] In this recipe, agency switches places several times. It begins with the beasts' presumed natural or original intention to attack people, then moves over to the natural desire of people to avoid being attacked. The desire to protect themselves results in the preparation of the paste to be carried to new places and burned as soon as a new residence is established. The final transition of agency goes back to the animals that will choose to avoid places in which the paste is burned, rewarding people with the forethought to anticipate marauding beasts and the knowledge to prepare recipes to discourage them. This kind of prevention becomes the province of female readers through a recipe that uses common ingredients and a familiar format to enable them to manipulate the external environment from within domestic space.

The second recipe in this section combines a welter of humoral remedies with advice to treat the evil actions of one person against another. It begins with the presumption that readers will be aware of the signs of poisoning, suggesting that this is a threat that they recognize as part of the hostile outside world. Once the poisoning is confirmed, all the recommended actions aim to remove the poison from the victim's body through as many channels and as quickly as possible. Recommendations for vomits, purges, glisters, and proper regimen follow, including having the victim drink olive oil, butter mixed with hot water, linseed, or nettles, all of which will push the poison out in both directions.[69] This wealth of medical advice, twelve recommendations in all, privileges the potential of readers to alter the hostile actions of the external environment. The recipe concludes, however, by placing agency in the hands of providence: "And in continuing with this meanes, he shall be deliuered with the help of God."[70]

The appearance of providence at this late date is not surprising in light of the complex role played by God in early modern English thought. The same God that produced enemies willing to poison you and gave them the ingredients with which to do it would help you to recover by providing the antidotes for the poison to those willing to look for them in nature, but he would save you only if you had suffered well.[71] The transition of agency from nature and humans to God in this recipe is not representative of the majority of entries in this book, which privilege human behavior and knowledge for their ability to control out-of-control natural forces, especially bodies. This recipe is different because it presumes the deliberate hostile action of one person against another.

That is the case for only one other recipe in the book, the one that tells readers how to remove iron from a wound. This entry does not mention God, instead leaving the healing of the wound entirely in the hands of the practitioner. In this case, skilled hands are required. The recipe urges readers, after the preparation of a valerian salve, to "then make your binding or band as it appertayneth [to the wound], by this means you shall draw out the yron."[72] The emphasis in this recipe seems to lie more on the application of the bandage than on the preparation of the relatively simple valerian ointment, but regardless of which step accomplishes more in the healing of the wound, both depend upon the knowledge and ability of the practitioner. Practitioners with some forethought are also rewarded by this remedy, which assumes that readers will have anticipated the possibility of a family member or friend being struck by an arrow or other metallic object and have either brought with them or will know the location of valerian root and materials for a bandage.

These three recipes paint a picture of an unpleasant, aggressive external world that can kill. They warn readers of the possible harm awaiting them outside their homes, especially in unfamiliar places or in the company of new people. Finally, they arm readers against potential harm, providing them with a short list of ingredients to keep close to hand and to replenish in new locations. These three recipes match the agency of readers with that of nature, in the shape of venomous beasts, and enemies, in the guise of poison and iron weapons. Although they do not always conclude with the reader in the position of total control, they do reveal the fine balance that can be wrought between readers and their environments through a bit of preparation, some skill, a lot of luck, and the help of God.

Conclusion

The three books examined here, while rather similar in context, stake out public and private spaces for women to use secrets, and they produce nuanced natural worlds that female readers and secrets might inhabit. The risks taken by Richard Jones, Edward Allde, and Edward White in printing these small books are acknowledged on their frontispieces and in their introductions, and their potential to make money had to outweigh the anxiety they provoked among consumers of cheap print. The strategies they adopted to evade that anxiety centered on space, with each text carefully grounding itself in spaces accepted as feminine, and then moving outward into more theoretical or intellectual arenas. All three of these larger projects buried their roots in the rich variety of tasks for which early modern women were responsible—the care of their own and their families' bodies, cookery, gardening, household decoration and maintenance, and caring for livestock.

Readers would have seen their lives mirrored in the range of problems for which these small books provided solutions, and they would have inherited from these texts a sense of their own ability to work from feminine spaces and familiar tasks to restore order to their frequently disordered worlds. They might have taken heart to see some answers to problems they hadn't yet experienced, or been pleased to add some new recipes to their collections kept handy to treat common medical problems. They might have even recognized some of the recipes in these texts: after all, Manus Christi and cherry conserve can only have so many iterations! They would have also seen the potential to make their worlds more pleasant, whether through careful spicing of foods, kind treatment of horses, the perfuming of gloves, or recipes for a range of amusements. All these factors unite these ultimately different texts into reflections of early modern feminine spaces, both mental and physical, and their relationships with secrets dedicated to the production and maintenance of natural order.

The natural world in these books is governable through the stepwise application of recipe ingredients and procedures. The consistent presence of recipes in all three of these texts signals the transition of unstructured, unpredictable natural processes into the realm of the structured and predictable. It also marks a particularly feminine approach to coping with nature, one that brings it inside and under control in the domestic space of the kitchen and even in the shape of the highly domesticated external space of the kitchen garden. The next chapter will discuss a masculine location and strategy for domesticating the natural world in the shape of an inexpensive book of stable and equine management for gentlemen and stable boys that will ground the care of animals in the production of elite masculinity.

CHAPTER FIVE

Secrets Bridled, Gentlemen Trained

Gervase Markham's horsemanship and horse care manual, *How to chuse, ride, traine, and diet, both Hunting-horses and running Horses. With all the secrets thereto belonging discovered: an Arte never here-to-fore written by any Author* served as a substantial part of Markham's empire of practical guides to animal husbandry and gentlemanly occupations. First printed in 1595, it was reissued intact and as parts of other books by Markham throughout the seventeenth century.[1] It represents Markham's first entry into the market of inexpensive print, and stands as part of a burgeoning collection of books by young gentlemen on the art of horsemanship. It is not the first to be published, but it is the only one to refer to the body of knowledge it addresses as a collection of secrets.

One focus of this chapter will be locating this text within the genre of books of secrets by analyzing the ways in which the word *secrets* works in the text, what it signifies, and how it shapes the highly practical language of the contents into a body of knowledge unified by a model of practice built upon an experientially determined theory. The second focus of this chapter will be the ways in which the book contributes to the definition of elite masculinity by allying it with the means and time to purchase and maintain expensive horses and enter them in competitions. The model of horse care presented here requires that the gentleman in question have sufficient funds to purchase well-bred stallions and mares and pay the salaries of the various people required to care for and train them. These people, from stable boys to stable managers, form the third focus of the chapter, on the ways in which the knowledge presented in the text was disseminated to those most likely to use it. Unlike the majority of inexpensive print, which is directed to readers of any class interested in purchasing a cheap collection of previously unavailable information, this book clearly delineates its purchasers as gentlemen and its audience as their servants. Markham's book insists upon attention to a set of questions that often elude careful analysis in other books in its price range, including the sharing of texts and necessary intersections between print culture and oral communication. It is the perfect book for examining the creation of a body of "secret" knowledge, the means of communicating that knowledge to two socioeconomically determined populations, and the extension of secrets beyond the workings of the natural world to inform both stable hands' work and a gentlemen's pastime.

The book has two explicit final products: a well-trained and well-maintained horse and a gentlemen who is a good horseman. The success of both horse and man, however, depends upon the book's implicit final product, a staff of skilled grooms whose lives are defined by the rhythms and demands of their work. Taken in combination, the activities that shape the lives of grooms, stable boys, and stable managers sketch the outlines of the equine industry in sixteenth-century England. The details of that industry have not attracted an enormous amount of historiographic attention, though English involvement in matters equine had its first upsurge under Henry VIII and played an integral part in the plans for his wars against France and Scotland. Joan Thirsk remains the authority on this topic, and her account of the place of horses in sixteenth- and seventeenth-century England, including a few paragraphs on Gervase Markham and his family's role in that history, appears in her broader work on England's rural economy at the time.[2] She argues that a combination of factors contributed to the rise of the horse and the ensuing improvement of equine quality in England, including the breeding program begun by Henry VIII in the 1530s and supported by the creation of his gentlemen pensioners, the steady increase in internal trade on roads and rivers that occurred between 1500 and 1700, the appearance of the coach as a means of elite travel, and the increasing popularity of mounted entertainment, including performances of classical military maneuvers, foxhunting, steeplechasing, and flat racing.

The breeding program instituted by Henry VIII and continued by Elizabeth I arguably played the central role in increasing the popularity and quality of horses in sixteenth-century England, since it encouraged the emergence of an elite class of gentlemen who valued and enjoyed producing, training, and playing with expensive, high-class equines. The breeding program was initially begun to reverse the losses of horseflesh from two decades of war with Scotland and France, and its goal was the production of horses of good size and conformation. The gentlemen pensioners, fifty men whose court duties included supplying horses for military and ceremonial occasions, were the first to be called in to address the shortage. They were each expected to keep at least five stallions of minimum size and quality, and they were rewarded with land formerly held by monasteries or captured in the wars.[3] In 1531 Henry made it illegal to export horses to Europe, and by 1532 it was illegal to sell horses to Scotland.[4] When the English supply of horses had been exhausted, Henry turned his attention to Europe, particularly Holland, whose equine population he soon exhausted. In 1544, for example, he requisitioned 9,600 horses from the Dutch, fewer than he wanted, and incited the wrath of the people of the Netherlands, who begged their queen regent to allow them to pay cash taxes rather than send more horses to Henry's wars. Their objections went unheeded, and the last sizable horses that could be located were rounded up and sent to the English.[5]

When Europe proved singularly unwilling to supply him with any more mounts, Henry resolved to improve the quality of English horses on a much larger scale than was possible with only the land and means of the gentlemen pensioners. In the process, he made the breeding and keeping of expensive horses an integral characteristic of rural gentlemen. From 1540 through the end of his reign, he required members of the nobility and other large landowners to keep stallions of at least fifteen hands (fourteen hands in the north) in their parks to improve the size and strength of English horses. In 1541 members of the nobility, gentry, and clergy were charged with keeping a number of quality stallions determined by their annual incomes, with dukes and archbishops required to maintain at least seven stallions who were at least three years old and stood fourteen hands high. The poorest gentry, defined as those with yearly incomes between five hundred marks and one hundred pounds, were required to keep at least one stallion of that age and size.[6] This policy, as well as the laws against exporting horses, was maintained through 1600, in order to rebuild the damage done to the English equine population by Henry's wars.

Elizabeth I was in many ways her father's daughter, and she shared his conviction that better English horses equated with a stronger English military and, as a result, a stronger English state. In 1580, in response to the war with Spain that she and her counselors perceived to be imminent, she created the Special Commission for the Increase and Breed of Horses. The commission ensured that gentlemen kept all the quality stallions they could and bred them as frequently as possible to mares of good lineage. It reinforced a new law requiring a minimum number of stallions and mares of appropriate size in gentlemen's parks and drafted a new law requiring that a stallion of at least fifteen hands be kept in every common area to ensure more progeny of good size. The gentry by this time, however, needed no laws to direct their attention to the production of nice horses. Many decades of royal encouragement had resulted in the widespread adoption of the crown's standards in the gentlemen's private breeding programs. Fine horses became a popular gift for one aristocrat to give another, since their appearance and strength proved a lasting testament to the wealth and prestige of the estate that produced them and improved the lines of the stable to which they were sent. The importance of the first point should not be underestimated, since horses who left their breeding stable as gifts kept the name of their breeder, so that Francis Walsingham, for example, would never forget that Pied Markham, one of his 102 horses, had been a gift from Robert Markham.[7]

The gentry also began to adopt the royal preference for elegant horses kept and ridden well, for pleasure as well as for military pursuits. The period of peace in the 1560s and 1570s reinforced Henry VIII's earlier practice of importing Spanish and Italian riding masters to train the horses in the royal stables and the gentlemen pensioners. During these years, younger sons of the gentry were frequently sent to Spain or Italy to learn the art of classical horsemanship, or European horse masters were imported to teach them at home.[8] Several of these young men responded to their training by translating European books on horsemanship or drafting their own.[9] The first printed book devoted to equines and written by an Englishman was by Nicholas Arnold, one of Henry's gentleman pensioners, who began importing horses from Flanders in 1546 and traveled to Italy for Edward VI to study Italian breeding and training programs. While his text is no longer extant, it was frequently cited by his contemporaries as an excellent guide to breeding military horses, and it was among the first to support the use of Neapolitan stallions. Thomas Blundeville wrote the next English equine text in 1560. It was a translation of the Italian book, *The Art of Riding*, by Federico Grisone, and had the high honor of being read in draft by William Cecil and cited by nearly every equestrian book that followed, including Gervase Markham's. Blundeville also wrote *The Fower Chiefest Offices Belonging to Horsemanshippe*, which saw five editions and a great number of reprints between 1565 and 1609. It corrected the mistakes Blundeville believed to permeate the Royal Stud at Tutbury, where violent domination rather than gentle persuasion prevailed as the preferred training method, and it provided a set of guidelines for improving stable design, breeding, and care, as well as more moderate training methods.[10]

Gervase Markham followed Blundeville, producing his first book on horses and horsemanship in the early 1560s. Thirsk asserts that Markham's books exemplified the new trend in English equestrian culture. "The new principles taught a professional attitude to horse training, making clear to what heights of perfection horses could be schooled in obedience to their riders. They emphasized the wisdom and utility of training horses by gentle persuasion, by sensitive use of the hand, rather than by cruel and harsh bullying or beating."[11] Markham continued Blundeville's tradition of writing a single book that included advice on all aspects of producing an equine athlete, including its breeding, feed, care, and training. He also followed Blundeville's lead in assuming that gentlemen interested in the production of fine horses would be both willing and able to teach their stable hands and other staff the fine points of his texts. While the books were certainly inexpensive enough to be purchased by people of small economic means, they were also clearly intended for consumption by men with the inclination and financial power to produce a stable of high-class pleasure and racing horses. The logical conclusion is that these gentlemen were then expected to direct their staff according to Markham's program.

The question remains, of course, how familiar the boys and men who cared for the gentry's stables would have been with horses, since they had a firm place in elite culture but were less firmly ensconced in the lower socioeconomic strata. Thirsk argues that the sixteenth century saw the rise of the horse not just among the elite, but at every level of society. She states that by 1558, long wagons were replacing single packhorses and carts as the preferred means of transporting goods and that each wagon required more horses to pull it than had the carts. She adds that an increase in river transportation made it necessary for more people to travel to ports, and horses became an increasingly popular method for going to and from home. She asserts that the 1550s were the first decade in which yeomen farmers were more likely to ride than walk, and that by the early 1600s, even peasants traveled astride.[12]

More frequent and more popular horse fairs also point to the increasing prevalence of horses in England. These sales featured all manner of horses—exquisitely bred and turned-out beasts intended to catch the eyes of royal and aristocratic buyers on the lookout for matched pairs to pull carts, foxhunters, and racehorses, and sturdy draft ponies destined to pull plows. Horse fairs became increasingly common and increasingly well attended in the 1560s and would draw increasingly large crowds from that time forward. In response to the growing number of people searching for horses, tenant farmers and freeholders improved their own small-scale breeding programs. Two examples provide evidence for the participation of horse breeders of all social classes in the trend to produce more and better English horses. In 1560 the Privy Council ordered all the local horse owners in the Lincolnshire fens to present their horses to the royal envoy, who would select the thirty best animals to be carriage horses for the court. The opportunity to present an animal, however, was not outside the range of possibility for an average farmer, who would have counted up to four horses among his animals.[13] Thomas Ogle whittled down the field to two groups of stallions and geldings and eventually purchased fourteen horses at Spaulding and sixteen at Boston, paying between two and four pounds for each animal, from a wide range of sellers.[14] This was not a negligible amount of money to a farming family. The breeding and sale of one high-quality horse, an option made more possible by the increasingly common presence of large, strong stallions on common grounds and the availability of horse fairs, would have been a good strategy for small farmers looking to increase their incomes.

While anyone with the land, funds, and patience could produce an elegant horse, the English obsession with equines was uniquely upper-class. In the early eighteenth century, Daniel Defoe would claim that the horses produced in the Vale of York, the area widely held to be supreme among English breeding grounds, were presented by their grooms at the peak of perfection, "they bring them out like pictures of horses, not a hair amiss."[15] The amount of time and energy needed to develop and maintain horses to that degree would have required a considerable number of grooms. Markham's text addresses the reader as though he alone would be doing the work of training, riding, racing, and managing his horses, but this glosses over the realities of the service workforce that performed the broad array of tasks covered in the book, including the building of a proper stable and pastures, the cleaning of that stable, horse breeding, classical military and race training, and the intensive daily care required by hunting and racehorses. When read with an eye to the people who would be doing the work described in this book, two kinds of knowledge transference are brought to light. The first involves oral communication between members of different classes, in this case between gentlemen and their servants.[16] Gentlemen may have been the primary purchasers of this book, since it provided them with advice on how to improve their breeding and training programs. Unlike the other books of secrets discussed in this work, however, the purchasers were not the ones who would be actually doing the work described in the text. This reifies the connection between the ownership of elite horses and gentlemanly status, since neither was possible without an extensive supply of devoted servants. The servants would have been the ones to make use of Markham's text, either as repeated in directions from their employers or through direct use of the book itself. Some servants might even have purchased their own copies of the book, particularly once it achieved popularity as a source of useful knowledge about raising and maintaining nice horses, in order to improve their grasp of equine matters, command a better position, or ask for a better salary.

The gentlemen who owned farms and had the means to purchase and maintain high-quality breeding, hunting, and racing horses also had the means to support a large number of hired hands to care for their stock.[17] These hired hands worked as grooms, as stable boys, as exercise riders, and in a variety of other positions dedicated to barn maintenance and the feeding, grooming, training, and reproduction of fine horses. Servants did a great deal of the labor in early modern England, and they were partially responsible for the success of agricultural ventures on both great estates and tiny, independent smallholdings. The culture of service dictated that young people of lower and middle social classes were sent out to work for other families in early adolescence and continued to do so until they had earned enough money to set out on their own and marry.[18] This saved their families the money required to support them and offered young men and women the opportunity to master a trade or animal husbandry and, in the case of young women, learn domestic skills like ale making and cookery. Young men were assigned the task of caring for large livestock, including draft animals, beef cattle, and sheep, while young women were more likely to be responsible for running the dairy, planting, weeding, and harvesting household gardens, and caring for small animals.[19] Servants were generally hired at job fairs, where they attached themselves to an employer through verbal or written contracts, most often for the period of a year. This combination resulted in a fluid population of people entering and leaving service, with only a small number identifying themselves as servants throughout their lives.[20] Service became a common way of entering adulthood, rather than a permanent socioeconomic indicator or profession, which implies that a wide range of individuals with various levels of formal education and literacy would have passed through the households of strangers and been exposed to their expectations for the best way to run a house or a farm, including a horse farm.

In his 1620 text *A Farewell to Husbandry*, Markham explicitly places horse care, including cleaning the stables three times per day, bedding stalls, grooming, and feeding, within the purview of a stable servant's chores.[21] He is less explicit in *How to chuse*. It is, however, hard to believe that he was speaking to gentlemen when he wrote, "for those three days [before a hunt race], you must night and day watch with him, making him to eate all the meate he eates, out of your hand," or in fact when he wrote any of the other time- and labor-intensive instructions in the book.[22] Gentlemen were meant to internalize his method for producing quality beasts and direct their servants accordingly. The amount of attention Markham lavishes on the proper care of these animals is not disproportionate to their role in determining social status in rural England. Horses occupied a unique place in farm culture; since they were more likely than other livestock to appear in public, their condition reflected the stability and quality of the yard that produced them. For the young boys whose lives were devoted to their care and the gentlemen who owned them, horses were a status symbol, and a well-bred, nicely turned out hunting, carriage, or racing horse brought glory to his owner and groom.

Masters and servants seem to have largely agreed upon the amount of care and effort that should be devoted to maintaining horses, probably since both benefited from their exquisite turnout, but occasional disagreements reflect the important place that horses occupied in the minds and lives of their caretakers. Ann Kussmaul notes that a few servants got into trouble for thieving corn and other feed intended for pigs and bullocks in an attempt to improve the condition of their equine charges.[23] The fact that grooms were willing to risk their positions to provide more and better feed for the horses under their care indicates that they were deeply invested in the potential for improving their own lots embodied by well-conditioned horses and that the cultural importance placed on fancy horses had direct relevance for the people responsible for their care.[24]

Gervase Markham's book was priced and edited in such a way as to make it as likely that it was read directly by grooms, since the gentlemen who invested fortunes in the name of improving English horses. The text would have informed their daily work and helped them to establish their own notions of good horse care. It could also have indirectly affected their work and expectations, as the gentlemen they worked for passed on Markham's advice on caring for and training horses. The young men who followed these instructions would have internalized them as thoroughly as if they had read them, if not more, and they would have been just as capable of taking this knowledge with them after they left service to start their own households and perhaps buy their first horse. The combination of processes through which this text would have circulated in stable yards suggests that it was important because it opened up the arena of horse care, an arena it designates as secret, to an audience of educated gentlemen and stable hands, and it privileges the knowledge, rather than the knower, as the most important demarcation of status.

In the world described by Markham's text, readers are defined as gentlemen in the first lines of the dedication, which reads "To the Gentlemen Readers." This is not as perfectly clear as it seems, since it opens up two possibilities for a more socially diverse readership. The first lies in the wording: If the dedication is addressed only to gentlemen readers, it does not preclude the presence of readers who are not gentlemen, it simply fails to address them. Servants would have been less likely to expect direct address anyway—the world of print was not yet a republic. The second is a bit trickier, since it redefines the word *gentlemen* by dissecting the phrase "gentlemen readers."[25] In the phrase, *readers* is the more important of the two words, which opens up the potential for all readers to be gentlemen. This interpretation, while requiring a bit more work than can have been considered standard for an average sixteenth-century reader, would have united all readers interested in breeding and maintaining expensive horseflesh under the heading of gentlemen. This, in turn, would have reinforced the connection between gentlemanly status and a preoccupation with horses.

Readers of all social classes would have been equally likely to read the dedication and interpret themselves as its intended recipients, even if a few stable lads had a laugh over the ridiculous nature of being addressed as gentlemen. If the phrase is interpreted in this way, then readership, and the possession of specific knowledge, assumes a social standing of its own. The title Markham bestows upon the readers who learn and follow his method for horse care is "wise," and he places them opposite "fools," particularly those claiming to be experts, whom he lambastes for endangering horses with their ignorance. Stable servants, whose presence in the text is defined by the fact that gentlemen would not have performed the heavy labor of building barns or cleaning stalls, are not denied the chance to be wise, unless Markham considers them to be as intrinsic a piece of his model as the expensive stallions he directs his readers to import. Since he gives no directions for the proper type and nature of servants to hire, aspects to which he devotes a great deal of attention when they relate to imported stallions, it seems likely that he is willing to admit anyone who knows his system entrée into the closed circle of those privileged by knowledge. This opens up the opportunity for everyone who practices the model to be both a "reader," if "reader" is the same thing as knower or practitioner, and a gentleman, since wise reader-practitioners are necessarily gentlemen. A great deal more than a clever textual analysis is at stake here.

The definition of an audience according to the secret knowledge it possesses rather than the social standing to which its members were born introduces the possibility of a much broader social purpose for books of secrets in sixteenth-century England. Rather than simply serving as cookbooks, medical recipe manuals, natural atlases, roadmaps to transmutation, or even learned debates about the characteristics and limits of art and nature, these texts also expanded terms like *wise, learned,* and *gentlemen* to accommodate every reader. While the arena of natural philosophical discourse would continue to be dominated by educated men, the importance of practical knowledge, including craft techniques associated with ink making, glass cutting, and even horse breeding, was on the rise, as were its practitioners. The process of knowledge dissemination from master to servants through oral communication raises two questions on the nature and content of knowledge transmission between people of different social classes, particularly when they are engaged in the same project only because one is in the employ of the other. First, how much of the language of secrets would have survived this kind of interclass communication, and what would it have looked like after translation? Second, what purpose did the language of secrets serve for this book, given that Markham would have been likely to anticipate its oral dissemination?[26] These questions, in turn, produce a third that interrogates the position of theoretical and practical knowledge within this book and, more broadly, in the genre of books of secrets. The first question is quite probably beyond the reach of historians, since it would have required a privileged position as eavesdropper among the stable lads of sixteenth-century England.[27] The remainder of this chapter will focus on the latter two questions, whose resolution lies not in the imaginary but in the language and in the lines of Markham's book.

Learning the Language of Secrets

The question of whether this text belongs in the genre of books of secrets has already been partially resolved by examining the language of secrets that dominates its discourse from the frontispiece to the printer's device gracing the last page with the words "The hidden truth becomes apparent in time."[28] The meaning of that language in this particular book has not yet been interrogated, and it deserves examination because it sheds more light on themes already examined in this work: the ways in which "secret" knowledge was understood and defined and the ways in which that knowledge could be revealed without losing its status as secret. The frontispiece, which heralds the first mention of the word secrets as well as the first appearance of the printer's device described above, introduces secrets in relation to the possibility of their discovery. That discovery is made possible by the text, which promises "all the secrets thereto belonging discovered" in its revelation of the fundamentals of horsemanship.[29] It is not inconsequential that secrets appear only in relation to their telling. The production and marketing of books containing secret knowledge depended upon the definition of some body of knowledge as secret. That process could be logical, as in the case of alchemy, or it could pull at the boundaries of logic, as in the case of cookery. Horse breeding, care, and training, because of their upper-class associations, would likely have fallen in the first category, but it could as easily have functioned in the second, since caring for horses frequently fell to servants in big houses, or as discussed in the last chapter, to women on smaller farms.

The trick of making secret something generally considered within the purview of common knowledge rests in separating the piece about to be "discovered" from the things that are already known. Markham performs this trick like a master, following the announcement of the secrets he will discover with the pronouncement that they comprise "an Arte never here-to-fore written by any Author."[30] He neatly divides equestrian knowledge into two pieces—one that is commonly known and widely believed to be complete, or at least sufficient, and one that has been so well preserved from prying eyes and ears that most people never even knew it existed. Writing plays an important part in this construction, since the lack of a written record of this body of knowledge is what permitted it to remain secret, at least to English readers who were not especially likely to have access to the Italian and Spanish books on the same subject. It is also central to the definition of knowledge as secret—according to this construction, knowledge that exists outside the confines of the written always can be secret, since it has never been published.

Writing and printing a book, especially a reasonably affordable one, should irrevocably change the nature of the knowledge it contains, since it moves it outside the small cluster of people who would have been able to access it through oral communication or in manuscript and makes it available to a much larger audience. Printing a book of secrets is akin to telling secrets, it is a process that seems to reduce the value of its contents by making them permanently and easily accessible. Fortunately for early modern printers, rather than adopting the cynic's stance and insisting that any book that has seen the light of day, particularly in English and for the measly price of four pence, is not worth having, early modern readers seem to have been willing to believe that information advertised as having been secret was desirable, particularly if it came from the hand of a gentleman who supported his text with anecdotes about how the material was obtained and gave his own experience as credibility. The secrets Markham told were no less valued by readers for their telling—they were, instead, valued for their prior status as secrets. Like the wonders of marigolds revealed again in *Albertus Magnus* and the secret to perfect quince tart unveiled in *The Treasurie of hidden Secrets*, the keys to feeding and caring for horses were also still considered to be secrets.

This trick, a second defining one of the genre, depended upon the division of the book's audience according to individuals' abilities to master the presented information and make use of it in appropriate ways. The characteristics of a good reader of secrets—discussed in other chapters, including diligence, patience, and carefulness—are reiterated throughout Markham's text in a process that constructs an ideal practitioner in parallel with the revelation of the promised secrets.

The frontispiece is, again, the place to begin discerning the construction of this audience. It announces that some of the information it presents will be entirely new and thus will constitute the secrets to be uncovered. It advertises the rest of its contents as an improvement upon existing texts and outlines the characteristics of a set of readers who would find it useful. "Also a discourse of horsemanship, wherein the breeding, and ryding of Horses for service, in a briefe manner, is more methodically sette downe then hath beene heretofore: with a more easie and direct course for the ignorant, to attaine to the said Arte or knowledge."[31] Secrets, once written, can be rewritten, and according to this, often needed to be in order for readers with little formal education or experience with alien epistemological systems to understand them.

The problem in producing good practitioners of secrets lay as well in the means of communicating the information—readers did not always possess the skills required to be a practitioner, and vice versa. Some readers could be made into good practitioners with the help of a revised edition that presented the knowledge in clearer language, using diagrams or illustrations, or with examples. Some readers, on the other hand, were never intended to be practitioners. Markham wrote his book for a group of men whose wealth and social position protected them from ever having to practice his labor-intensive system for producing hunt- and race-ready horses. They were, instead, supposed to understand it well enough to convey it to the men in their employ who would then be responsible for all the physical aspects of horse maintenance. They needed an accessible, manageable edition of Markham's advice so that they could convey its fundamental pieces to the largely uneducated and highly fluid group of young men who worked as their servants. These young men could have come into service with their own ideas about the best ways to maintain horses, gained from their families, former employers (including those who may have been reading Markham), and other servants. Their ability to either put aside their own convictions or, in the case of a young man whose experience meshed with Markham's system as interpreted by his master, improve upon the model he was expected to follow, would determine his success as a practitioner. In this case, the definition of a good reader is almost certainly not the same as the definition of a good practitioner.

Good readers in Markham's world match those described in other texts in their ability to internalize his directions and see how they fit into a larger model for producing better, stronger, and faster horses, but in this case they are notable for their ability to translate those directions into manageable pieces to be performed by people who might have little or no knowledge of the bigger project to which their labor contributes. Good readers are good translators, and good practitioners are good listeners. The best practitioners might have been those who were able to grasp the outlines of the larger model from the pieces they were asked to perform. More likely, they were those young men who had been in the service of equestrian gentlemen for long enough to have been exposed to multiple pieces of the system and understand how they contributed to and benefited from the creation of more nearly perfect horses capable of winning races and bringing glory home to their grooms and their owners. The glory to be gained from having been the caretaker of a good horse was not insubstantial. Since standard service contracts ran for the period of a year and then were either renegotiated or ended with the servant going back to the job fair to find another position, a year full of successful horses in your charge had direct financial benefits. Young men could either command higher wages from their current employers because they had contributed to their masters' successes on the track and in the breeding shed—since their departure would threaten that winning streak—or they could go back on the job market asking for higher wages because of their achievements in their last position.[32] Their internalization and consistent, effective use of Markham's system would remove it from the print culture of secrets to the broader culture of artisanal knowledge, where it could be communicated orally among practitioners of different levels and elaborated upon according to practical experience.

The possibility of contributing to a solution for the vexed audience question does not justify assigning a text to the books of secrets genre. The case for considering Markham's book and its occupation-directed compatriots like *The Seamans Secrets* and *The Boke of Secrets* derives from the language of the text and the multiple meanings of secrets in circulation in sixteenth-century England. Four sets of similarities can be derived from a comparison of Gervase Markham's book to other books of secrets discussed in this project, namely *The Mirror of Alchimy*, *The Book of Secrets of Albertus Magnus*, *The Treasurie of commodious Conceits, and hidden Secrets*, and *The Treasurie of hidden Secrets*. First, the categories that Markham works with match those displayed in all these books. It is perhaps most like *Albertus Magnus* in this sense, since both books display a shared heritage of Renaissance naturalism. For example, the book of marvels in *Albertus Magnus* announces, "And not withstanding everie thing hath his owne naturall vertues, by which everie thinge is a beginning of a mervailous effect."[33] Markham presents his adherence to the unchanging basic natures of beasts in his argument for a particular pasture design. He asserts that the enclosure for foaling mares should include a fresh spring, "my reason is, because it is the nature and property of Mares, to covet to foale eyther in the water, or as neere as they can possibly gette."[34] These two excerpts reinforce the similarity between the two books in their assertion that knowledge of the intrinsic virtues and characteristics of natural things will benefit someone who hopes to make use of them. Both texts assert that art will necessarily fall short of its potential when it is applied without reference to the essential natures of the materials on which it is being exercised.

A shared conception among books of secrets of the composition of natural things and the importance of that composition for the practice of art establishes the next similarity between Markham's text and other books of secrets. All these books address the process of using art to improve and extend the reach of natural products or beings, but perhaps none does it so evidently as *The Mirror of Alchimy*, which includes three treatises that elaborate the arduous path one must follow to create the philosopher's stone and perfect base metals. The language Markham uses to describe the effects of training on the raw material of horseflesh bears marked similarities to that used to describe the wonders wrought on nature by the alchemical art. The first line of the final treatise in *The Mirror of Alchimy* asserts, "although Nature be mightie and marvailous, yet Art using Nature for an instrument, is more powerfull then naturall vertue, as it is to be seene in many things."[35] Markham ascribes to a similar view of the relationship between art and nature, although the natural objects he strives to improve with his art are the bodies and natures of horses. In his discourse on starting a colt under saddle, he urges the rider to raise his hands "that you may gather uppe his necke to the uttermost height, that arte or nature wil by any meanes suffer it."[36] *The Mirror of Alchimy* and *How to chuse* also share an enthusiasm for finding the most difficult subjects and considering them a challenge or a test for their arts. The alchemical treatise attributed to Roger Bacon exclaims, "These things are almost as much as nature or Art are able to performe. But yet the last decree, wherein the perfection of Art can doo ought with all the power of nature, is the prolonging of life for a great space, and the possibilitie hereof is approved by many experiments."[37] In his section on the training of young horses, Markham writes of the test posed for his art by the most limited and argumentative beasts, stating, "Neverthelesse, if the horse be dull and stubborne, of which kind I most intreate, for in them is the depth of art to be tried."[38]

Gervase Markham's text shares another aspect with books of secrets engaged in more evidently natural philosophical projects. All these books present evidence in similar ways, producing a series of recognized experts to support the conclusions of the author, who remains the primary source of authority in the text and reserves the right to augment or even correct information that has been presented previously as authentic. *The Secrets of Albertus Magnus* exemplifies this approach through its consistent citation of ancient authority followed by an iteration of the "author's" own conclusions in order to establish the text's place in the tradition of natural philosophical knowledge. The book begins by citing Aristotle in support of its primary intellectual project, claiming "Aristotle the Prince of Philosophers saieth in manie places, that every science is of the kinde of good things" and then completes the introduction with the author's thoughts on the subject, "Of the which saying, two thinges are concluded."[39] Markham calls on experts from Xenophon to his own contemporaries, particularly Blundell, to supplement his conclusions in his equestrian manual. He writes at the end of his recommendations for breeding horses, "For other things which I have omitted, as touching the speciall markes of Horses, theyr complexion and colours, theyr sundry kinds, theyr natures & dispositions, I refer you to Grison or Blundell, who of those things have writ sufficiently."[40]

In the end, however, he presents his own experiences with breeding and training horses as the ultimate evidence for his words. In the introduction to his veterinary treatise, for example, he writes, "For mine owne part, my intent is to write nothing more then mine owne experience, and what I have approved in horses disease most available."⁴¹ He even offers his experience as evidence that other authorities are incorrect and deserve to be abandoned, an approach that permeates the treatment of differing authorities in several of the books of secrets already analyzed here. In the first treatise of *The Mirror of Alchimy*, for example, the author exclaims against prevailing methods to create the philosopher's stone, writing "O extreame madness! What, I pray you, constraines you to seeke to perfect the foresaide things by strange melacholicall and fantasticall regiments?" and suggests his own, naturally derived method instead.⁴² In his discussion of the proper way to introduce mares to stallions for breeding, Markham argues against the prevailing wisdom of "Xenophon, Vegetus, Grison, and all our English writers" and asks, "If therefore my reasons and practice shall be found in equall balance with theyr former judgments, I doubt not but the censures of the wiser will allow me, though the ignorant carpe at my wrythings."⁴³ He asks his readers to be the arbiters of his conclusions, grounding them in his years of experience, which he offers up as equally compelling evidence to counter the inherited knowledge of venerable authorities.

Markham's book also shares with other books of secrets an explicit set of expectations for readers, specifically that they be intelligent, diligent, and devoted to their art. All these books adopt a didactic tone that engages readers in the process of learning the secrets of nature outside the traditional academic locations of the university and the guild. In the world of books of secrets, readers' personal characteristics, rather than a formal education, will allow them to master the information they encounter. Furthermore, these books argue that readers characterized by their intelligence, diligence, and devotion will be better masters of nature than will their more formally educated peers. In *The Mirror of Alchimy* and Markham's text, the early sections of the books determine desired characteristics for readers. The title treatise in *The Mirror of Alchimy*, for example, makes evident its expectations for readers when it announces "That which hath been spoken, everie Alchemist must diligently observe."⁴⁴ It also makes evident the benefits to be gained by a careful reader: "Be therefore wise: for if thou shalt be subtile and wittie in my Chapters (wherein by manifest profe I have laid open the matter of the stone easie to be knowne) thou shalt taste of that delightfull thing, wherein the whole intention of the Philosophers is placed."⁴⁵ Markham makes clear in several examples the high standards of dedication he expects from his readers. In the first section of his text, he delineates his ideal horse farm, including the construction plans for three separate paddocks with individual water sources and highly varied terrain, as well as the plans for a safe and snug stable. He anticipates some concerns that readers might raise when his suggestions conflict with prevailing ideas about proper housing and care of mares and foals, but he does not hesitate to believe that his readers, once convinced that he is correct, will follow his instructions to the letter and make use of his text to plan and build their farms, as well as care for their charges. To do so would have required careful reading and diligent practice accompanied by tremendous expenditure.[46]

Markham also makes more direct remarks about the ways in which he hopes his readers will absorb and use the information he presents. He begins his treatise on breaking colts with instructions for haltering them and adds, "which I would wish to be done with all the gentleness and quiet means that may be."[47] The importance of gentle and quiet manners when dealing with horses is repeated throughout the text, and it is paired with an equal emphasis on persistence. In his directions for teaching a young horse to lead, he writes about the importance of repeating a reprimand until the horse stops disobeying: "and faile not but as oft as he strives to breake away, so oft do you pluck him backe with these suddaine straines and twitches."[48] He summarizes his idea of good horsemanship in his advice for tacking up a young horse:

"Let his keeper be alwaies trifling and doing somewhat about him, eyther rubbing or clawing him in one place or other where he shall finde him most ticklish or daintie: still giving him kind words, as ho boy, ho boy, or holla love, so my nagge, and such like termes, till he have won him to his will that he will suffer him to dresse him: take uppe his legge and picke him in every place: provided always, (and let both his Rider and Keeper hold it as an especiall rule of good horsemanship,) never to do any thing about a Colte, eyther suddainly, hastily, or rudely."[49]

All these books base the importance of following their instructions on the potential consequences of poor workmanship. Markham stresses the disasters that can result when horsemen fail to attend to his advice to be persevering and gentle, consistent and kind. "For when unfaithfull Horsmen wil come to their Horses with suddaine motions, and violent furies, that makes Horses learne to strike, to byte, to starte at the saddle, to refuse the bridle, and to finde boggards at mens faces."[50] The consequences of failing to follow the instructions in *The Mirror of Alchimy* were just as disastrous, if not as physically painful, since they would have required starting again at the beginning of the arduous and lengthy process of making the philosopher's stone.

Alchemy and horsemanship were also similar in the financial and physical demands they placed on their disciples, since both required the careful, consistent manipulation of expensive raw materials that could easily be ruined through inappropriate or careless application of these arts. Following these arts correctly required a tremendous level of discipline, a trait that was stressed as desirable in texts devoted to explicating both traditions. The two arts also shared a dual notion of progress, in which the person practicing the art was improved by the process of improving the natural ingredients with which he worked. In the case of Markham's manual of horsemanship, the end result was not just a classically trained mount or a competitive racehorse, but also a horseman who realized the virtues of starting with high-quality horseflesh, as well as the importance of patience and perseverance in training it to its highest potential.

The process of reading and performing the prescribed routines to develop a young horse into a valuable riding animal also transformed a young rider into a valuable horse trainer, a man skilled in a classical art.[51] Gervase Markham remarks indirectly on this dual project by repeatedly stressing the importance of patient repetition and the learning potential that difficult horses offered. Rather than urging his readers to purchase trained mounts who could educate them by responding only to correct commands, or easily schooled horses, who would rapidly reward correct handling and training, Markham praises horses who are slow to learn. "Note that if in a dayes riding or two, ne three, you can not bring him to that perfection you woulde, that then you be not discouraged, but continue your labour, for those Horses that are the slowest of conceite, and hardest to understande theyr Riders meaning, being once brought to know what they must do, are alwaies the surest holders, and ever after, the perfectest performers of any lesson, whatsoever."[52] This excerpt develops the theme of the multiple benefits to be gained from mastering the art of horsemanship, particularly in the context of developing a difficult horse. Following Markham's model would, in the end, produce a man who could consistently produce nicely behaved and fully trained horses, a consummate trainer.

The Mirror of Alchimy, in the same manner, promised not just a single act of transmutation but the creation of an individual who understood and could steadily improve all base metals and also manipulate other natural processes to his own benefit. Both the base metal and the alchemist, in fact, would be advanced by following the steps outlined in these treatises, and only those men who had forsaken self-indulgence and entirely devoted themselves to studying the book of nature would be rewarded with control over the natural world. Horsemanship and alchemy share standing as traditions of secrets, and both disciplines spawned books of secrets that required readers to wholeheartedly devote themselves to learning the art in question. Only through complete devotion would men become alchemists or horsemen, masters of the natural world.

Markham's text bears a final resemblance to other, more recognizable books of secrets: It, like Albertus Magnus's collection of secrets and John Partridge's multiple books of medical and culinary recipes, offers cures for common health problems in recipe form. Two similarities unite the veterinary section of this book with the medical and veterinary recipes offered in *The Treasurie of hidden Secrets* and the images of the natural world presented in *Cornucopiae, Or divers secrets*. The first can be found in the introduction, which aligns Markham with a particular theory of disease and gives directions for the process of restoring a body to health. He writes, "In generall, sicknes is an opposite foe to nature, warring against the agents of the body and minde, seeking to confounde those actions which upholde and maintaine the bodies strength and livelihood," then refers readers with greater interest in the topic to other equine authorities, particularly Vegesius Rusius and Blundeville.[53] This definition draws on the Hippocratic theory of illness, which dominated early modern thought and identified disease as the result of individual humoral imbalance and medicine as the active intervention required to correct that imbalance. Broader explanations and explorations of the place of systems of sympathy and antipathy in the natural world filled books of secrets, as has already been demonstrated in the preceding chapters. These books share an affection for a world in which seemingly inexplicable phenomena can be explained through careful observation of natural virtues and the interactions between and among natural objects, and they dedicate concentrated effort to decoding the systems of sympathy and antipathy that link seemingly distinct elements in the natural world. Books of secrets also favor a style of presentation that encapsulates and makes accessible those interactions, the recipe.

Recipes appeared in books devoted to everything from cookery to natural magic, and they also formed the core of individual collections that contained an individual's or even several generations' attempts at making predictable and manageable an often hostile and unpleasant natural world. Recipes were so popular because they compressed the complex webs of interactions in nature into short, easily recalled and repeated instructions that presented both the required ingredients and the desired end products of specific manipulations of natural objects. Two kinds of recipes appear in Gervase Markham: The first is immediately recognizable and part of a nicely defined tradition of recipes, while the second is less accessible and more integrally entwined with the textual project, like the textual structure in *Cornucopiae*. The first type includes the recipes for veterinary interventions, which always appear in relation to the description of the problem they are intended to correct and bear a distinct resemblance to the medical recipes in texts like John Partridge's culinary and medical recipe books for women, the veterinary advice in *The Widowes treasure*, and the magical or natural philosophical recipes in *The Book of Secrets of Albertus Magnus*. Markham, for example, gives the signs that suggest that a horse has a fever, accuses farriers and most farmers of misdiagnosing these symptoms as bots or bewitchment, and then provides a cure for the disease. The unifying characteristic among the recipes in the veterinary section of Markham's book and those presented in both Partridge's texts and the text attributed to Albertus Magnus is their clear definition of the problem at hand and the appropriate ingredients and steps to be taken to solve it. Each recipe is devoted to a specific aspect of curing disease and takes advantage of an intricate system of natural relationships to produce the desired result. Taken together, these recipes outline the understanding of the natural world that produced them. In other words, an entire book of recipes reflects the interactions, relationships, and sympathies that, once mastered, allow people to navigate and control the world, not just the specific problems that they are intended to solve, and not just the world within the text.

The second type of recipe included in Markham's book has a very different purpose. Rather than introducing a system for managing the natural world that extends beyond the confines of the text, it defines the two most integral pieces of its textual world, namely the horseman and the horse. Seventy-six of the 136 pages in the text are devoted to the production of a skilled horseman and a schooled horse, and they represent extended recipes for the creation of these two products. Creating a horseman, according to this text, requires the basic ingredients of a young, privileged man with the time, funds, diligence, and patience required to devote himself to learning the art. It also requires a course of instruction that will steer potential equestrians away from harsh, inappropriate, and hasty training techniques and direct them toward a gentle, correct, and patient regimen. Finally, it requires devotion to the multiple aspects of horsemanship that accompany riding, including knowing about bloodlines, breeding, and the proper care of broodmares, stallions, and foals, managing illnesses and injuries, and feeding, grooming, and exercising horses appropriately for the jobs you want them to do. This book provides that course of instruction, thus defining the parameters of a good horseman while making itself indispensable to the production of these individuals.

The book intertwines its definitions of a good horsemen and a good horse, embedding the production of the former in its instructions for the creation and maintenance of the latter. It once again renders itself indispensable by creating a new definition for the perfect equine, then giving the directions for creating it, and in so doing, defining the parameters of a skilled horseman. The recipe for the ideal mount begins with its creation, which revolves around Markham's argument for basing a breeding program around the Arabian courser, a breed that had penetrated Spanish and Italian breeding but had not yet made much of an impression on the English.[54] After describing and dismissing as inadequate a plethora of known and appreciated bloodlines, Markham introduces his choice of the ideal breeding stallion. "Now to come to the true Stallion, who for his brave trotte and pure vertue of valure in the fielde, is a staine to all other Horses: whose comelie and easie amble, may be an eternall instruction to all Aldermens Hackneys, howe to rocke theyr Maisters into a sound sleepe, whose wonderfull speede both in short and long courses, may make our English Prickers hold their best runner but Baffles, who by nature hath all things perfect, nothing defective."[55]

The next ingredient in the recipe for a properly trained horse is correct handling in its early years, before it is old enough to be gentled. Markham is particularly clear about the proper timing and management of weaning. "Your foales having run with your Mares the space of a yeere, or within a Moneth, in so much that they are readie to foale againe, I would wish you to draw them from their Dams, and lock them in some close house for a night: then in the morning take them, and to give each of them two or three sippes of Saven, and so to let them rest two or three houres after: this Saven is a most soveraigne Medicine for the wormes, which will be most abundant in young foales insomuch that if they have not present remedie upon the first drawing from their Dammes, they wil many times suddainly drop away and die."[56] Two crucial aspects of producing a quality horse are made clear in this section. The first is the proper timing of weaning, which is designed to separate foals from broodmares when the former are old enough to survive by grazing rather than nursing, and ensures that broodmares can be bred as frequently as biologically possible. The second piece points to the wide variety of knowledge required to produce a good young horse, as it involves correct veterinary management of the weaning process. It is not enough, according to this text, to be a good rider and trainer, you must also be knowledgeable about the veterinary needs and proper care and management of your horses.

The portion of the text that introduces Markham's model for preparing field hunters and race horses highlights the importance of proper care and management. The book devotes more pages (37) to the housing, feeding, grooming, and regimen for hunting and racing horses than it does to instructions for breaking and training them (33), which reflects the weight that maintenance and care have in the equestrian world imagined in this text. If the point is not made solely by the pages devoted to these concerns, Markham speaks about it directly: "I will therefore, not make them two artes, but one, making this latter [dieting and management] an apendex to the former [training], concluding him (in my foolish judgment) not an absolute horseman, which hath not understanding in them both."[57] The value given to the quotidian duties of horse care suggests that men who were interested in producing competitive horses would do better to learn the fundamental aspects of good care than to depend on the knowledge of stable managers, grooms, and stable boys. The latter group might be inclined to follow the instructions of farriers and leeches, "who cure many times what they know not, and kill where they might cure, knew they the cause," or follow the tradition that claims foxhunting a sure path to lameness, rather than being acquainted with Markham's credo, "that Hunting horses are never lamed through theyr immoderate riding or labour, if they have a good keeper."[58] This text was not directed toward an audience of stable boys, but rather the gentlemen who hired them. The information it contained would not have stopped with its readers, however, because it was designed to be passed on from newly minted skilled horsemen to the people who performed the hard labor of maintaining, feeding, and grooming expensive horses.

The book elaborates upon, corrects, and clarifies an existing knowledge tradition. It appropriates that tradition and makes it accessible to an audience of horse owners, predominantly men of wealth and social standing, who could adapt it for use by their stable hands. They would, in turn, appropriate it through repeated use and make its contents so well known that it would enter the realm of protected artisanal knowledge even while it was commonly known. This trajectory is supported by the absence of the word secrets from the title pages and leaves of (as far as I can tell) every book on horsemanship published in early modern England after 1596, including Markham's own additions to this text and other books on the subject. In combination with the increasing importance of the equestrian industry in the seventeenth century, the knowledge that supported the creation of quality horses ready to run and win races became increasingly widespread and integral to a growing field.

This little book was so successful at telling equestrian secrets for two reasons. First, it was exceptionally good at advertising itself—its frontispiece announces the revelation of an entirely new body of knowledge, the clarification of existing texts, and a new treatise addressing and resolving an old topic. Second, the structure of the text and the language Markham uses to present his information is both clear and accessible. There is no hidden subtext to search for, no extremely close reading required. The art of horsemanship is laid bare in a series of steps that lead from breeding a quality animal through weaning, halter breaking, and training. It is then extended to include the correct preparation of horses for hunts and races, both in the field and in the stable yard. The puzzle is completed by the promise of cures to the diseases that had long plagued horses' vulnerable bodies. The importance of the veterinary section of the text is made apparent in the table of contents. The breeding, riding, and training of horses take up 4 chapters, while the cures for a wide variety of equine diseases are the focus of the next 61. This is made possible by the devotion of a chapter to each disease, a common technique probably adopted from human medical books, and more pages of the text are devoted to the first 4 chapters than to any of the remedies. The fact that each disease is listed in the table of contents is important—Markham knew that horses were fragile and prone to acquiring strange diseases with frightening names and supernatural overtones, like "moon eyes," "the sleeping evil," and "the staggers." He offers a text that familiarizes readers with these diseases by clarifying their causes and their symptoms and providing a manageable, relatively easily administered cure.[59]

The existence of these diseases in print, rather than in the darkness of a stall or the mind of an itinerant farrier, contributed to their manageability. As soon as a set of symptoms could be assigned to a category and given a title, you had a horse "that is taken" rather than one "bereft of his feeling, moving, or stirring," an explanation, and a cure.[60] You also knew what to avoid and how to proceed. Markham is clear in his dismissal of occult explanations of equine diseases, and he introduces his position in his justification of training and dieting horses to be field hunters. He argues that this kind of regimen cannot be considered a risk to horses' health because the causes of equine diseases are clearly understood and are neither present in nor encouraged by his dieting program. He also asserts that a horse that falls sick will always be recoverable by a skilled human practitioner because his disorder is natural, and therefore curable through the correct application of medical remedies. "First, there is no disease nor infirmitie in a Horse, especially within his heade or bodie, which be secrete and unseene, and therefore most dangerous and mortall, but a man skillfull in his arte, shal both discerne it before it come to extremitie, and also recure it."[61]

The problem in this model is not the horse, nor the disease, but unskilled, overconfident practitioners who use ill-considered methods to treat disorders they do not understand. In his chapter on horses who are taken he writes, "yet some Farriers, not well understanding the grounde of the disease, conster the word taken, to be striken by some Plannet or evill spirit, which is false."[62] This is a clever manipulation of the genre, since it calms the reader while attributing the fears he might have had to a group that many owners might consult. Furthermore, it eliminates that group as serious competition for Markham's text by dismissing its members as ignorant and greedy. Farriers bear the brunt of Markham's wrath in the veterinary section. He describes them as ignorant fools whose superstitious and ill-conceived practices bear little resemblance to the humorally derived recipes presented in the text. He also places his own experience with using his recipes over the claims to expertise made by farriers, who lack the longevity dealing with equine diseases to support their claims to expertise.

Markham makes the most of this comparison, measuring his own successes in comparison to the clumsy failures of lesser practitioners and remarking upon his allegiance to the successful coupling of theory and experience as opposed to fear and superstition. The first recipe in the veterinary section is an excellent example of his tactics and appears to have been carefully chosen, since it cures a disease that Markham claims most practitioners fail to acknowledge as even possible in horses and therefore frequently misdiagnose. He begins with a description of his experience with the disease, writing, "yet I have my selfe seene of late, (both by demonstrate opinions of those better learned and by the effects of the disease) some two horses which I dare avouch were mightily tormented by a fever."[63] Note that he does not stand alone, even in this announcement of his experience. He asserts the confirmation of his diagnosis by experts even before he presents the symptoms that led him to that diagnosis, so that his interpretation of the case is never isolated from external opinion nor permitted to stand without support. His attacks on the practitioners who failed to correctly diagnose the disease are justified by his own success, particularly as his success is supported by the weight of men with more experience and education. He announces, "though divers Leeches had thereof given divers opinions, one saying it was the bots by reason of his immoderate languishment: another affirmed him to be betwicht, by reason of his great shaking, heaviness, and sweating: but I have founde it and approved it to be a feaver, both in effect, nature, qualitie."[64]

He proceeds to give the humoral nature of the disease and recommends a course of treatment that corrects the surfeit of bad blood through bleeding and sweating.[65] He also describes the correct regimen for maintaining the horse while he is recovering, including giving him water in which mallow, sorrel, and purslaine have been boiled, and only sodden barley, with occasional handfuls of rye, for feed. He concludes the section with another testament to the importance of theory-driven experience in this section of the text: "Divers have written of divers Agues, and I coulde prescribe receipts for them, but sithence I have not been experimented in them all, I mean to omit them, intending not to exceede mine owne knowledge in anything."[66] His reaffirmation of the truth of experience is not simply a policy statement, but a rationale for leaving out recipes with which he has no experience. This strategy also serves to reiterate his distance from the itinerant practitioners who willingly treat diseases they believe to have hidden causes and no known cures.

The veterinary section is not the only one to praise experience, nor is it the only one to assert Markham as different from other practitioners, including accepted authorities. The language he uses to assert his points is remarkably clear and comprehensible, and he makes a practice of announcing when and how he is departing from accepted wisdom. His first real departure takes place in the very beginning of the book, when he discusses the best location for raising and keeping horses. He recommends a high plateau and writes, "Nowe it may be objected to mee by some, that I wronged my selfe in chusing of high groundes, sith they be neither so fruitfull of grasse, nor so convenient for water as lower grounds be."[67] Having listed both the objections and the rationale behind them, he addresses these concerns. "But my answere is, hee that breedes upon low groundes that be fruitfull and full of ranck grasse, and keepes his Mares onely for breede and not for worke, shall finde by proofe, (as I have done) that in the winter season when they shall come to standing, the most of his race Mares, especially those which goe over, shall hazard to die of the rott."[68] The evidence he presents for his departure from standard practice is based on his own experience with a disease that had long been considered rampant in sheep and not particularly detrimental to horses. He announces the loss of some of his best mares, those bred to race and purchased for their ability and their potential for producing the next generation of fine English horses, as a sad and expensive lesson in following inherited wisdom, and he proffers his solution as an inconvenient but effective means for preventing this disease from taking hold in another herd. He also offers a list of the additional benefits to be gained from moving horses to higher ground, combining his own experience with the well-known Hippocratic axiom that high, dry places with fresh water have the lowest rate of epidemic diseases and their inhabitants enjoy better health.[69]

He follows this pattern of signposting his departures from common practice, listing potential objections, and providing his rationale for choosing a different path, in the first four treatises. It is a practical approach, since it clearly delineates how and in what ways Markham is different, it establishes the novelty of his method, it repeatedly asserts the importance of personal experience in the creation of his method, and it allows readers to make their own decisions about how to proceed. It also supplies them with a rationale for choosing a new approach to building a stable yard, breeding a colt, or training a young horse: Rather than leaving readers to announce that they have departed from the wisdom of Xenophon to follow Markham, the text supplies them with the tools required to make and defend their choice. This is an effective strategy for spreading the system to a wider audience, since it opens up a space for debate over the text's contents that invites more people to read it, even if they do so only to counter its approach. Markham's style of presentation is sufficiently self-confident and his evidence, supported by anecdotes and claims of personal experience, sufficiently compelling that the text promotes itself.

Part of this successful strategy depends upon the seamless way in which the first four chapters fit together. Markham leaves little room for the reader to take one piece from his model and combine it with others. He presents every part of his work, from the building of a good breeding farm to the training of a classical military horse or a successful racehorse, as the inevitable product of having accomplished the preceding step. He begins his section on breeding, for example, "When you have therefore your groundes severed and used as I have before written, it then resteth that you stock it with Mares, which for their strayne, collour, and comely shape, will be profitable to breede upon."[70] He presents an entire system of good horse care and a model of the estate on which to do it, complete with illustrations in the first section of the text demonstrating the proper design of a shelter for a foal paddock.[71]

The world created by Markham's text has a clearly delineated list of problems and challenges as diverse as stubborn, irascible colts that resist training, broodmares with less than ideal confirmation and a range of diseases. The author accumulated details of all these disasters from his own experience. The same experience supports a series of responses that have been proven to prevent, control, or resolve these hazards. Thus, the problems are presented alongside their solutions, and both exist to support the general model of the world proposed in the text. In this model, most dangers are known and manageable, and readers willing to follow Markham's directions in their own yards are banking on the dependability of his experience to warn them of impending threats and guide them through the best ways to negotiate them. Readers are not left without hope, however, should they encounter an entirely new problem. Having invested their faith and a great deal of money in another man's experience, they can turn back to the text and generalize his approach into a system for addressing similar problems that will reflect the principles Markham has forwarded as integral to the structure of the natural world.

Under these conditions, Markham's little book moves from a model for a stable to a model for problem solving in husbandry that depends upon the application of existing theoretical systems like humoral medicine, Hippocratic ideas about the environment, and Italian classical horsemanship, to personal experience to produce a classically based solution to individual problems. The only part of the text that lacks allusion to classical treatises is Markham's treatise on the dieting and preparation of hunting and racing horses. This entry in the text is an elaboration of existing practices for producing and maintaining performance horses and lacks classical allusions because it caters to uniquely early modern interests. The secrets Markham promises are his own inventions, and they form only one piece of the puzzle he presents in this text. Perfect knowledge of the system and mastery of the ancient traditions and contemporary experience that inform it permit the best use of his intricate and intensive equine improvement program. Good readers, who in their turn can become and produce good practitioners, are capable of achieving that level of mastery through careful study of the book.

Conclusion

The English would come to dominate the disciplines of steeplechasing, fox-hunting, and flat racing in Europe, and the assumed alliance of the English upper class with equestrian culture was well established by the eighteenth century. Gervase Markham's book marks an important moment in the transition of the equestrian arts from a European body of knowledge to a remarkably English one. It conveys all the information that the English had purchased, absorbed, and borrowed from visiting Spanish and Italian horse masters and their own experiences abroad. It locates their experiences within that tradition and uses them to shape it to fit English circumstances. Finally, it extends that tradition in a direction determined by English equestrian interests and English terrain to produce a kind of horsemanship and a kind of horses that best suited their burgeoning desire to be horse people. Markham's book is particularly interesting because it points to a moment in which a body of knowledge that would soon come to be common was still in the hands of a privileged few. It represents an attempt to take the breeding standards created by Henry VIII and improved by Elizabeth I, standards designed to be met by a few noblemen on their impressive estates, and make them a truly national standard.

By 1596 the demand for high-quality horses was encouraging the expansion of their production outside the limits of the aristocracy to everyone with enough land and money to have a brood mare. The nature and eventual purpose of those horses is evidently not military but instead represents a new focus on elegance and speed. Markham recommended importing Arabian bloodlines to improve English stock. Arabians are a notoriously small breed, known more for their elegance, stamina, and speed than their size. Markham promoted the creation of English horses who could meet the demand for light, fast animals who could win flat races or, when crossed with native English lines, compete in the steeplechase. He also wanted to contribute to the population of gentlemen's field hunters, whose worth was increasing as their discipline developed and more men of means participated in foxhunting. His book marks the moment when established equestrian knowledge met an Englishman's agenda for the future of the English horse, advertised as a secret, and integral to the creation of a manageable, predictable, and inevitably equestrian world.

His book is also important as an example of sixteenth-century print culture and exceptional as a demonstration of intersections between print and oral communication. While other books advertise their contents to audiences "of the meanest capacities" or "ladies," Markham's intentionally blurs the socioeconomic boundaries of his desired audience by directing his text to gentlemen but filling it with work that would be accomplished by servants. In doing so, he sketches a much more complete picture of the relationship between gentlemen and their servants, and gentlemen and their estates, than is made possible by less explicit texts that strive for a broad audience by aiming low in their pronouncements of desired reader characteristics, or attempt to prove their exclusivity by defining an elite reader population that they could not possibly have expected to exclusively command.

Markham creates a textual world that mimics the social relationships of the English estate, where gentlemen were expected to direct their servants' work according to the constraints of a grand plan derived either from ancient authorities, contemporary experts, or with increasing frequency, a combination of both. The role played by servants involved, of course, unswerving obedience to their masters' instructions and a general willingness to proceed with or without access to the larger plan under construction. Access to that plan would likely have been determined by seniority and centrality on the estate—older men with several years of horse care under their belts would have been more likely to be integrally involved in the realization of the longer-term breeding, maintenance, and training project than would young boys of little experience who were expected to leave after a single year of service. The interactions between these groups were undoubtedly more complex than these typologies indicate. Ann Kussmaul reports that some servants were punished for their independent reallocation of livestock feed to their equine charges, and while she does not interpret this behavior beyond noting the glory to be gained from fit and attractive horses, she opens the door for some consideration of the multiple expectations that gentlemen and servants had for their interactions with horses and their relationships with each other.[72] The minutiae of those relationships are quite probably inaccessible, but the mandate given by Markham of communicating the elements of a theoretical model through practical directions given by gentlemen to servants suggests that "readership" and access to printed knowledge are far from the same thing.

The last significant thing to be taken from this book is its relationship to the word *secrets*. Its language and format nicely situates it within the body of books dedicated to revealing the secrets of nature in inexpensive, vernacular language editions. It resembles more predictably located books of secrets in its descriptions of desired reader characteristics, like diligence, perseverance, and carefulness, in its combination of theoretical and practical language, and in its design for a predictable, manageable world. It departs from more classical books of secrets, like *The Secrets of Albertus Magnus*, which seek to reveal the complicated interactions among various natural objects and place them under human command, because it focuses entirely on a single topic. It constructs its model of the world around the creation, maintenance, and perfection of those objects, rather than from a complex web of interactions whose totality remains inaccessible through the vantage point of individual entries. It also departs from other single-subject books of secrets, like *The Mirror of Alchimy*, because it clearly presents its model for the mastery of a particular process, rather than encoding it in a morass of tightly coded language that demands devoted bouts of intensive reading. Finally, it departs from cookbooks, like *The Treasure of hidden Secrets*, in that it presents remedies derived from a single medical theory, humoralism, as proved by a single person's experience, which could, but did not have to be, confirmed by other authorities.

Markham's text extends the world of secrets to a knowledge tradition that arrived in England at royal bequest in the heads and hands of Spanish and Italian riding masters who passed it on to gentlemen and their sons who trained at the royal stables or traveled abroad. The body of knowledge itself lies between natural philosophy and craft, since it is not simply a mechanical system for producing artificial objects, but a theoretically determined practical model for manipulating and perfecting natural objects. Like alchemy, the breeding, training, and maintenance of horses constitute a coherent model for consistently improving a naturally occurring object or class of objects by following a consistent program for controlling the passage of it or them through processes derived from those found in the natural world. Like alchemy, horsemanship and its accoutrements are occupations of privilege. But the breeding, care, and training of horses also bears a certain resemblance to craft knowledge. Like ink making, there is money to be made from the creation of well-bred, well-trained horses. The labor of equine creation and maintenance is performed by young men whose responsibilities and knowledge of the entire system increase with their seniority in the yard, and more especially, in the trade. Markham's text fills the space between books of secrets devoted strictly to uncovering and offering plans for mastering the relationships between objects in nature and those dedicated to the protection of trade secrets that formed the base of the English mercantile economy. It describes an occupation of gentlemen made possible by the labor of servants, and it lays the groundwork for the assimilation of that occupation across boundaries of wealth and position.

CONCLUSION

A Secret by Any Other Name

The books of secrets analyzed here reflect, perhaps more than other forms of cheap print that remained popular for a longer period of time, their cultural and intellectual context. While the knowledge they contained was collected from a variety of sources and presented in the name of different audiences and disciplines, they depict remarkably coherent and unified images of the natural world that represent concerted efforts by compilers and printers to satisfy prevailing questions about the power of art over nature, appropriate uses and users of art, and the systems driving nature and their susceptibility to human manipulation.

I have tried to demonstrate that books of secrets share responses to these questions that include the creation of natural worlds composed of natural systems, particularly sympathy and antipathy, that are available for manipulation by wise and careful readers. Each book presents the possibility of mastery over its construction of nature through the diligent study of these systems, and they are careful to reiterate the importance of selfless effort in the pursuit of natural domination. The devoted readers imagined in the introductory letters and admonitions in these books, however, are quite different from the readers conjured by the questions the books strive to answer, which reflect baser human desires, such as the ability to command the love and loyalty of another, to detect lies or to lie with impunity, to fool adversaries, and even to conquer armies.

The books' descriptions of appropriate readers and practitioners of secrets as private individuals pursuing knowledge to better understand their place in the natural world and in divine order also conflict with the highly public nature of their production and sale and the potential for their public use. The division between public and private command of natural knowledge in the sixteenth century was more charged for some readers, particularly women and the uneducated, than for others. A desire to understand and attempt to influence the natural world was an acceptable thing for a gentleman, but less acceptable for a gentlewoman, and problematic for women and men of common means. Books of secrets, which promise reserved knowledge to a broad range of readers for a small price, erode the boundaries between aristocrat and layperson, male and female, even while their introductory letters reiterate the importance of those divisions. Their consistent presence in the midst of structuring binaries indicates the ineffectiveness of categories like "elite" and "popular" to describe knowledge made available in inexpensive editions, and it points to the presence of a highly mobile, shared system for understanding the natural world and the place of humans in it.

Inexpensive books of secrets provided readers from a wide variety of backgrounds with models of the natural world that were vulnerable to human manipulation. They shaped readers' perceptions of the places they occupied in the world, and they centralized, or at least acknowledged, readers' desires and offered directions for actualizing them through the manipulation of natural forces. They revealed, for everyone to read, the secrets of nature. So why did they lose their standing in the marketplace in the seventeenth century?

I do not believe that they were no longer needed—the importance of making predictable and malleable worlds accessible to large parts of the population would not have lessened in the atmosphere of cultural and political instability that characterized much of the seventeenth-century English experience. Instead, I would argue that they were replaced by two genres of books that accomplished very similar goals but were not strictly devoted to the revelation of secrets: books of knowledge, with which books of secrets had coexisted throughout the sixteenth century, and closet books, which took up an increasingly large part of the cheap print market after 1600.

Books of knowledge, inexpensive collections of treatises on astrology, physiognomy, and medicine, were directed more toward making the natural world predictable and explicable than exposing it as vulnerable to human manipulation. These books, perhaps best exemplified by the series of perpetual almanacs and prophecies that began to be attributed to Erra Pater in 1535, provided explanations for things that had already happened, delineated methods for determining the future, and, in the case of the medical sections, gave general advice for preventing the onset of disease, as well as specific treatments for physical ailments. Their intention was to give readers a map of signs by which to navigate their way through the world, reading the faces of strangers, the heavens, and their own bodies as pages upon which nature's script had already been written.

These books change the readers' position in relation to the natural world from the one of possible power spelled out in books of secrets to one of reactor, a relation wherein the readers are always striving to make sense of the things that nature has brought into their lives. Nowhere do these books recognize readers' desires for anything but the knowledge to read the book of nature in order to improve their chances of survival and success, and they do not offer readers the chance to make any difference in the forces affecting the world around them. The genre title, books of knowledge, reflects their approach: These books promise to provide readers with the epistemologies required to understand, and sometimes predict, nature; they do not promise to reveal the forces directing the natural world or provide recipes for manipulating them to "readers of the meanest capacities."

Books of knowledge, perpetual almanacs, yearly almanacs, wonder tales, and prophecies filled popular demand for explanations of the ways in which the natural world worked while offering readers instructions for predicting it and structuring their lives accordingly. Closet books, which flourished in the seventeenth century, provided a different center around which readers could structure and make sense of their lives: cultural expectations and personal conduct. The connection between books of secrets and closet books is difficult to miss. As early as 1573, Richard Jones printed *The Treasurie of commodious Conceits, and hidden Secrets and may be called, The Huswives closet, of healthfull provision*, signaling the shared domain of these two words. Books of secrets were limited by the dimensions of "secrets" to the dissemination of natural knowledge and the practices that produced and manipulated it. Closet books could extend further, governing conduct in the kitchen and prayer closet, the farmyard and the sickroom.

These books were united, however, in their project of making formerly private knowledge and behavior part of the public domain, and in their effort to provide a set of models for domestic, medical, and religious conduct. Books of secrets and closet books also share the task of making knowledge previously reserved for academics and aristocrats available to a broad audience. Books of secrets opened up the possibility of natural manipulation to a range of readers from a variety of academic and economic backgrounds. Closet books, on the other hand, opened up the possibility of cultural mobility for these same readers by making available to them guidelines for conduct in a variety of situations. They operate as a genre to reflect the continuing importance of the boundaries between the public and the private in the seventeenth century, and they highlight the importance of gender and its relationship to conduct in both private and public contexts.

The question of gender as a determining factor in defining appropriate knowledge, delineating appropriate behavior, and qualifying readership is another element that continues to inform inexpensive print in the seventeenth century. The closet books aimed at female readers reflect continuity of women's areas of expertise and areas of the household in which they held sway, particularly the kitchen, the sickroom, and the garden. They also reflect the ongoing importance of the concerns spelled out in the introductory letters to the three sixteenth-century books of secrets aimed at female readers, including the limits of appropriate knowledge for women, the boundaries around female use of natural knowledge, and an increasing emphasis on the importance of female readers conforming to standards of culturally determined conduct. This last point is complicated, because as scholars like Sasha Roberts have illustrated, female reading was not accepted unconditionally as a positive habit.[1] The content of inexpensive books aimed at women became increasingly geared toward their areas of expertise, particularly the management of their own and family members' bodies, and was less likely to place women in positions of control, or even prediction, in relation to the broader natural world. One of the questions raised by the transition from books of secrets, which accomplished the latter goal, to books of knowledge like *Aristotle's Masterpiece* and conduct books, is how women's knowledge became increasingly related to their physicality, whether as disordered bodies requiring medical management or disorderly bodies requiring social management.

While books aimed at women proliferated in the seventeenth century, they were not necessarily read exclusively by women, and they were not the only texts that women read. The dreaded "readership question," which has haunted early modern print scholarship since its conception, proves less resistant to analysis in the face of a broader understanding of the audiences to which different books were directed as interpreted from their introductory letters, their prices, the language in which they were printed, and the kinds of knowledge they contained. Print scholarship also benefits from considering books in their own context, by placing them next to similar books by similar printers, and attempting to locate them within the print marketplace. Efforts like these are beginning to be commonplace in print history, but this book opens up a new avenue for efforts to make sense of early modern print and its readership: the theater.

Printed editions of dramatic texts, particularly in quarto form, were both affordable and accessible to a wide variety of readers. The language of the majority of early English dramas, at least those that played on the public stages, was English, and, perhaps more important, sufficiently colloquial to invite readers into the action rather than scaring them away. Finally, dramas reflect prevailing national and cultural concerns that also informed other forms of inexpensive print, although not always as obviously. A careful reading of the natural worlds written for the stage and the characters who were capable of, or failed to, manipulate them provides insight into the ways in which nature, natural knowledge, and those who tried to alter the course of the world around them were constructed during this period. A comparative reading of sixteenth- and early seventeenth-century manipulators of nature, ranging from Faustus and Friar Bacon to Oberon, for example, suggests the different cultural positions these characters could occupy and makes those positions available to the people who attended theater, an audience similar the one reading and buying cheap print.

A focus on theater, both in its staged form and as printed text, also reflects the continuing importance of oral communication for knowledge transmission in early modern England. As Adam Fox and Tessa Watt have so aptly pointed out, oral and print culture coexisted and reinforced each other, rather than competing, and text often had multiple iterations, rather than existing solely as a single, silent encounter between a reader and a book.[2] These two forms of communication built up the number of people for whom print, or at least the knowledge it contained, had meaning, and they expanded the range of printed materials by making them a subject of family and friendly exchange. The possibility of incorporating dramatic scholarship into investigations of inexpensive print promises to support these claims by making cheap books part of a larger cultural exchange about questions like the ways in which the natural world worked, the appropriate limits of human knowledge about and manipulation of nature, and proper justification for testing the boundaries of art.

It has been a pleasure thinking about these questions and the ways in which they were expressed, argued over, and packaged in small books printed in England in the second half of the sixteenth century. Books of secrets, even more than four hundred years after their printing, remain powerful tools for bringing the natural world to heel and making it accessible and vulnerable to manipulation by a wide range of readers. They also, even after all these years, keep their secrets by raising more questions than they answer and remaining charmingly resistant to essentialist notions about natural knowledge, print culture, and early modern readers.

NOTES

Introduction

Many of the books studied here have unwieldy titles. To complicate matters, several of the chapters compare books with very similar unwieldy titles. To simplify references to them, the full form of each title is given only in the Bibliography. Please consult the Bibliography for the full identity and source of each primary source cited herein in short form only.

Pagination style varies in the primary sources. Whenever the source used numerical pagination, the Notes reflect that. A few sources used folio pagination, in which case the citation consists of the folio letter, the page within the folio, and the side of the page as a superscript r or v (for recto—right-hand side, or verso—left-hand side). For example, a citation to $H4^v$ refers to folio H, page 4, on the left-hand side.

1. Terry Pratchett, *Hogfather* (New York: HarperCollins, 1999), 392.

2. William Eamon, *Science and the Secrets of Nature: Books of Secrets in Medieval and Early Modern Culture* (Princeton, N.J.: Princeton University Press, 1994), 6–7.

3. Thomas Kuhn, *The Structure of Scientific Revolutions* (Chicago: University of Chicago Press, 1996).

4. Eamon, *Science and the Secrets of Nature*, 5.

5. William Warde, trans., *The Secretes of the reverende maister Alexis of Piemonte* (London: n.p. 1566).

6. Ibid., 9.

7. Pamela O. Long, *Openness, Secrecy, Authorship: Technical Arts and the Culture of Knowledge from Antiquity to the Renaissance* (Baltimore: Johns Hopkins University Press, 2001), esp. ch. 3.

8. Adam Fox, *Oral and Literate Culture in Early Modern England, 1500–1700* (Oxford, U.K.: Oxford University Press, 2000); Markham, *How to chuse*.

9. *Cornucopiae*.

10. Steven Shapin, *A Social History of Truth: Civility and Science in Seventeenth-Century England* (Chicago: University of Chicago Press, 1994), chapters 2 and 3.

11. Ibid.

12. Bernard Capp, "Separate Domains? Women and Authority in Early Modern England," in *The Experience of Authority in Early Modern England*, edited by Paul Griffiths, Adam Fox, and Steve Hindle, 117–45 (New York: St. Martin's Press, 1996).

13. Steven Orgel, "What Is a Text?" in *The Authentic Shakespeare and Other Problems of the Early Modern Stage*, edited by Steven Orgel (New York: Routledge, 2002), 1–2. See also Stephen Greenblatt, "Toward the Poetics of Culture," in *The New Historicism*, edited by H. Aram Veeser (New York: Routledge, 1989), 1–12.

14. Lorraine Daston, "The Nature of Nature in Early Modern Europe," *Configurations* 6 (1998): 149–72.

15. Richard Helgerson, *Forms of Nationhood: The Elizabethan Writing of England* (Chicago: University of Chicago Press, 1992), 6.

16. Patricia Crawford and Laura Gowing, *Women's Worlds in Seventeenth Century England* (New York: Routledge, 2000); Sara Mendelson and Patricia Crawford, *Women in Early Modern England, 1550–1720* (Oxford, U.K.: Oxford University Press, 1998)—see especially the introduction for the struggle to solve this historical problem; Laura Gowing, *Domestic Dangers: Women, Words, and Sex in Early Modern London* (Oxford, U.K.: Clarendon Press, 1996); Patricia Crawford, *Women and Religion in England, 1500–1720* (New York: Routledge, 1993); Sara Mendelson, *The Mental World of Stuart Women: Three Studies* (Brighton, U.K.: Harvester Press, 1987).

17. For examples of this kind of work, see Sara Pennell, "Pots and Pans History: The Material Culture of the Kitchen in Early Modern England," *Journal of Design History* 11,3 (1998): 201–16, Jennifer K. Stine, "Opening Closets: The Discovery of Household Medicine in Early Modern England," Ph.D. dissertation, Oxford University; and Margaret Pelling, "The Women of the Family? Speculations around Early Modern British Physicians," *Social History of Medicine* 8,3 (1995): 383–401.

Chapter 1: Printing Secrets

1. Arber, Edward, ed., *A Transcript of the Registers of the Company of Stationers of London, 1554–1640* (Gloucester, Mass.: Peter Smith), 1: xxxiii.

2. John King received licenses or was fined for printing without a license the following books in 1557–58: *A nose-gaye/the schole howse of women; A sacke full of newes; The Defense of women; Adam Bell &c; A jeste of Sir Gawayne; the Boke of Carvynge and Sewynge; Sir Lamwell; The Boke of Cokerye; The Boke of nurture for mens sarvauntes; The brevyat chronicle of the kynges* (in octav.). In 1558–59 he was fined for printing *The nutbrowne mayde* without a license. In 1559–60 he was fined for printing *Salomons proverbs* without a license and then licensed the same book. In 1560 he was licensed to print *Lucas Urialis nyce wanton; Impaciens poverte; The proude wyves pater-noster; The squyre of Low degree; Syr Deggre; Juventus;* and *Albertus Magnus*. In 1560–61 he was licensed to print *Lubsettes workes; The little herbal; The greate herbal;* and *The medysine for horses* (Arber, 1: 75, 79, 93, 126, 128, 150, 151, and 153).

3. Arber, 1: 46–47.

4. Arber, 1: 49–50.

5. In the year 1557–1558 he printed three books addressing aspects of cookery and

household conduct, namely *The Boke of Carvynge and Sewynge, The Boke of Cokerye,* and *The Boke of nurture for mens sarvauntes.* The first two books would appear, in pieces, in Edward Allde's edition of both books of the *Good Huswives Jewel,* which would also contribute to Richard Jones's editions of Partridge, *Treasurie of commodious Conceits.* These connections will be explored later in this chapter. For the licensing information on John King's books, see Arber, 1: 79.

6. Arber, 1: 41.

7. Arber does not date Copeland's signature, as he does for the members who appear on the earliest possible charter of the company. We know from Duff that he was printing as early as 1548 at the shop started by his deceased father, so that he would have been a likely candidate for signing an early, if not the earliest, edition of the charter (Arber, 1: xxxiii; Edward Gordon Duff, *A Century of the English Book Trade Short Notices of All Printers, Book-Binders, and Others Connected with It from the Issue of the First Dated Book in 1457 to the Incorporation of the Company of Stationers in 1557* [London: Printed for the Bibliographical Society by Blades, East, and Blades, 1905], 32–33).

8. Duff, 32–33.

9. Arber, 1: 48–50.

10. Duff, 32–33.

11. Mary's close watch over printers and her willingness to retract printer privileges will be shown again and in better detail in the section on William Seres, the man who printed *The Boke of secretes of Albertus Magnus* in 1570 and who was directly affected by Mary's policies.

12. Arber, 1: 93, 96, 101.

13. Thomas Dawson later printed from the Three Cranes at the Vinetree, and he produced *The Seamans Secrets* from that address in 1599 (Duff, 32–33.).

14. He printed *An introduction to knowledge* in 1562–63, *A breaf and pleasuante treatese of the interpretation of Dreames* in 1566–67, and the book of wonders in 1567–68. If his first two books of magical or natural philosophical knowledge, namely *The Book of Virgil* and *An introduction to Knowledge* sold relatively well, it is not surprising that he chose to print *Albertus Magnus,* especially since he continued to display interest in producing books on natural knowledge and the permutations of the natural world until the end of his career. For printing dates, see Arber, 1: 211, 339, 340.

15. Arber, 1: 392.

16. Elizabeth reinstated this privilege to William Seres in a letter dated June 3, 1559. In a later letter, she added that this privilege should be passed on to Seres's heirs. For a full copy of the first letter from Queen Elizabeth, see Arber, 2: 60–61.

17. This transfer had to have constituted more than equipment, since Denham began to use one of Seres's devices as well, first as his assign and then later claiming it as his own. For details of the movements of this device and the transfer of copies from Seres to Denham, see R. B. McKerrow, *Printers' and Publishers' Devices in England and Scotland, 1485–1640* (London: Printed for the Bibliographical Society at the Chiswick Press, 1913), 145, 181.

18. This summary is drawn from a report written by Christopher Barker on the

printing patents granted between 1558 and 1582 and found in the Burghley papers in the British Museum. It is included in an effort to contextualize sixteenth-century print culture in Arber, 1: 114–16.

19. Arber, 1: 50 and 62.

20. William Seres's date of death is not perfectly known but is alluded to in a note in the apprentice records of the Stationers' Company, in which "Mistres Seres late wife of William Seres" presents Robert Robinson for his "making free" (completion of his apprenticeship) between the 24th of February and the 27th of June 1580. Seres was followed into the printing business by his son of the same name, who appears in the *Registers* as William Seres, junior (Arber, 1: 682).

21. Arber, 1: 269.

22. Arber, 1: 356, 359.

23. Arber, 1: 364.

24. *The Secrets of Albertus Magnus* was not among the copies that Seres leased to Denham, nor did Jaggard inherit any of Denham's business after his death, and so the connection between these three men does not seem to affect the production of this book (McKerrow, *Printers' and Publishers' Devices*, 88).

25. Arber, 2:126, 672; McKerrow, *Printers' and Publishers' Devices*, 88.

26. This is not the first connection to be found between Jaggard and Gervase Markham's books. In 1607 Jaggard printed Markham's *Cavelarice, or The English Horseman: Contayning all of the arte of horse-manship, as much as is necessary for any man to understand, whether he be horse-trader, horse-ryder, horse-hunter, horse-runner, horse-ambler, horse-farrier, horse-keeper, coachman, smith, or sadler. Together, with the discovery of the subtill trade or mistery of horse coursers, & an explanation of the excellency of a horses understanding, or how to teach them to doe trickes like Bankes his Curtall: And that may be made to draw drie-foot like a Hound: Secrets before unpublished, & now carefully set down for the profit of this whole nation* (London: Printed by E. Allde and W. Jaggard for E. White, 1607). More will be said about Edward Allde and Edward White, who also teamed up to print *The Widowes treasure*, a book of secrets for women, which will be discussed later in this chapter.

27. For example, he printed Thomas Hill's *School of skill, containing two bookes: the first, of the sphere, of heaven, of the stares, of their orbes, and of the earth &c: the second of the sphericall elements, of the celestiall circles, and of their uses &c* in 1599; Edward Topsell's *Historie of Four-Footed Beastes: describing the true and lively figure of every beast, with a discourse of their severall names, conditions, kindes, and virtues . . . collected out of the volumes of Conradus Gesner and others* in 1607; Gervase Markham's *Cavelarice; or, The English Horseman* in 1607; Edward Topsell's *Histore of Serpents; or, The Second Booke of Living Creatures* in 1608; Sir Walter Raleigh's *History of the world* in 1621; and Richard Allestree's *New almanacke and prognostication for the yeere of our Lord God, 1621* in 1621.

28. These claims must be considered in the broader context of the books in which they were made and the state of the discipline. Neither Margaret Spufford in her *Small Books and Pleasant Histories* (Athens: University of Georgia Press, 1982) nor Tessa Watt in her *Cheap Print and Popular Piety* (Cambridge, U.K.: Cambridge University

Press, 1991) can be accused of a lack of historicism in their attempts to describe the intended audience for inexpensive print, particularly because they were among the first historians to notice that there were inexpensive books in the early modern period. They have provided us with an excellent basis from which to advance in our descriptions of sixteenth- and seventeenth-century English print.

29. Julian Roberts and Andrew G. Watson. eds., *John Dee's Library Catalogue* (London: London Bibliographical Society, 1990).

30. R. J. Fehrenbach, ed., *Private Libraries in Renaissance England: A Collection and Catalogue of Tudor and Early Stuart Book-lists* (Binghamton: Medieval and Renaissance Texts and Studies, 1992).

31. The best evidence I can find suggests that this was, indeed, the first English edition of this book. The English Short Title Catalog refers to a possible 1586 edition printed by G. Robinson for Edward White, but the original, which is in the British Library, lacks a frontispiece and has the title *The Widowes Treasure* and the date 1639 handwritten below the heading of the letter to the reader. I cannot confirm that this edition appeared before the Allde/White collaboration in 1588, since neither version appears in the Stationers' Registers; and, since there was a 1639 edition of this book printed by R. Badger for Robert Bird, I have chosen to treat the Allde/White version as the earliest English printing. It is worth noting that the Robinson/White and Allde/White versions are identical, to the extent that they can be compared without a frontispiece for the former.

32. Arber, 1: 2, 3.

33. Examples of each of these genres, all from Arber, in chronological order are *Ye tragecall comodye of Damonde and Pethyas*, 1567–68 (1: 354); *A manifest or a playne Dyscourse of a hole packefull of popysshe knavery*, 1569–70 (1: 416); *A pamphlet in the praise of Capt. Frobisher in forme of A farewell at his Third voyage in Maye 1578 by the northeast seas toward the Iland of Catea*, 1578 (2: 327); *A voyage by master Cesar Frederick merchant of Venys into the Easte India and Indyes and beyond the Indyes & c*, 1588 (2: 492); *A miraculous and monstrouse but moste true and certen discourse of a woman (nowe to be seene in London) of the age of 60 yeres in the middest of whose forehead by the wonderfull worke of God, there groweth out a Croked horne of 4 ynches longe*, 1588 (2: 504)—Edward White also licensed this, along with Thomas Orwin and Henry Carre—*Articles of household discipline, conteyninge precepts and prohibitions meete for a chrystian familye*, 1590 (2: 543) and *Certen instructions observaccons and orders militarie requisite for all Chieftains, Captaines, and higher and lower men of charge*, 1594 (2: 647).

34. *Aristotle's Masterpiece* was a popular guide to pregnancy and childbirth that first appeared in England in the late seventeenth century and quickly spread to America. Mary Fissell, "Hairy Women and Naked Truths: Gender and the Politics of Knowledge in *Aristotle's Masterpiece*," *William and Mary Quarterly* 60,1 (2003): 43–74, discusses this kind of misappropriation of *Aristotle's Masterpiece*—or the possibility for multiple appropriations of it.

35. *A ffarynge for mayde and wyf* and *The comly behavyour for ladyes and gentlewomen*, Arber, 1: 439, 442.

36. Arber, 1: 302.
37. *A glasse for vayneglorious women, conteyninge an envictyve againste the fantasticall Devises in womens apparrell*, in Arber, 1: 669.
38. Arber, 2: 324.
39. Arber, 2: 567.
40. *The right rule of Christian chastitie: profitable to bee read of all godly and vertuous youths of both sexe, bee they gentlemen or gentlewomen, or of inferiour state, whatsoever/collected and written by one studious to gratifie his friends, and profit his kindred*, in Arber, 2: 358.
41. Arber, 2: 404.
42. Arber, 2: 543.
43. *A tru description of a chylde borne with Ruffes in the parryshe of Myttcham in the County of Surry*, Arber, 1: 329.
44. Arber, 1: 33, and McKerrow, *Printers' and Publishers' Devices*, 4–5.
45. Arber, 2: 308.
46. The fifteen were *The Arbor of Amytye*; *Turbervilles songes and sonnettes*; *The merrie meeting of Maides*; *Newes from Nynyve*; *The castell of Christian*; *A Ryche storehouse for gentlemen*; *The greene forreste*; *The fourthe Tragedie of Seneca*; *Newes out of Paules Churchyard*; *Palmistrye*; *The pityefull state of the tyme present*; *Hilles Physiognomye*; *The Travelled Pilgrym*; *A Contemplacon of misteries: Morrall Philosophie* (Arber, 2: 359).
47. *A miraculous and monstrouse but moste true and certen discourse* (Arber, 2: 504).
48. Arber, 2: 652.
49. McKerrow, *Printers' and Publishers' Devices*, 175.
50. *Praelectiones in duodecim prophetas minores—The lectures or daily sermons of John Calvin, upon the prophet Jonas, by N. B. [Nathaniel Baxter] whereunto is annexed an excellent exposition of the two last epistles of S. John, done in Latine by . . . August Marorate and Englished by the same N. B. . . .* (London: by John Charlewood for Edward White, 1578).
51. *The geomancie of Maister Christopher Cotton, gentleman: a booke, no less pleasant and recreative, then of a wittie invention to know all things, past, present, and to come: whereunto is annexed the wheel of Pythagoras, translated out of French into our English tongue* (London: John Wolfe, 1591), and *Philomela: the lady Fitzwaters Nightingale*, by Robert Greene (London: Printed by Robert Bourne for Edward White, 1592).
52. For a terrific explication of this tradition, see the introduction and chapter 1 of Long, *Openness, Secrecy, Authorship*, and the introduction to Eamon, *Science and the Secrets of Nature*.
53. It is, of course, possible that the lack of a record simply means that White was never caught for illegally printing this title, but the connections reviewed above between the two men, particularly in producing this text and the other books of secrets they produced in the name of a female audience, make an agreement between the two men about this title seem likely.
54. *Of prayer and meditation: contayning foure-teene meditations, for the seaven*

deayes of the Weeke: both for mornings and evenings: treating of the principall matters and holy misteries of our fayth, by Luis de Granada (London: Printed by James Roberts for Edward White, 1602).

55. *Cavelarice, or The English Horseman* includes much of the same information as Markham's first book of equestrian secrets, and its title makes the same promise of secrets theretofore unpublished. It came out after the wave of first-edition English books of secrets had ended and is noteworthy for its reworking of previously printed material as well as the information it adds to thinking about changing definitions of the word *secrets*.

56. The two titles for women were *A Christiall glasse for Christian women: containing a most excellent discourse of the godlye life and Christian death of Mistress Katherine Stubbes, who departed this life in Burton upon Trent in Staffordshire, the 14 day of December: with a most heavenly confession of the Christian faith which she made a little before her departure, as also a most wonderfull combat betwixt Sathan and her soule, worthy to be imprinted in letters of golde and to be engraved in the table of every Christian heart, set downe worde for worde as she spake it, as neere as could be gathered by Philip Stubbes, gent.* (London: Edward Allde for Edward White, 1603), and *The French garden for English ladyes and gentlewomen to walke in; or, A summer dayes labor: being an instruction for the attaining unto the knowledge of the French tongue, wherein for the practise thereof are framed thirteen dialogues in French and English, concerning divers matters from the rising in the morning till bedtime: also the historie of the centurion mentioned in the Gospell, in French verses: which is an easier and shorter methode then hath yet beene set fourth, to bring lovers of the French tongue to perfection of the same, by Peter Erondell, professor of the same language* (London: Printed by Edward Allde for E. White, 1605).

57. His first play, *The Comical history of Alphonsus, king of Aragon*, was printed in 1599 but performed in the early 1590s, and his later works included *The Tragicall historie of Friar Bacon and Friar Bongay*, first performed before 1592, and *George a Pinner of Wakefield*. For biographical sketches of Robert Greene, see Charles Read Baskerville, ed., *Elizabethan and Stuart Plays* (New York: Henry Holt, 1934), 247–48, and William Allan Neilson, ed., *Chief Elizabethan Dramatists* (New York: Houghton Mifflin, 1911), 855, 869–70.

58. The myth of the English vagrant counterculture and its production, as well as the realities of this social problem, can be found in Paola Pugliati, *Beggary and Theatre in Early Modern England* (Aldershot, U.K.: Ashgate, 2003), and A. L. Beier, *Masterless Men: The Vagrancy Problem in England, 1550–1640* (London: Methuen, 1985).

59. The others include Thomas Dekker, *Satiro-mastix; or, The untrussing of the Humorous Poet. As it hath been presented publikely, by the Right Honorable, the Lord Chamberlaine his Servants, and privately, by the Children of St Pauls* (London: Edward White, 1602); Michael Drayton, *The owle* (London: Printed by Edward Allde for Edward White and Nicholas Ling, 1604); and N. B. [Nathaniel Baxter], *Sir Philip Sydneys Ourania: that is, Endimions song and tragedie, containing all philosophie* (London: Printed by Edward Allde for Edward White, 1606).

60. Robert Greene, *Mamilia: A mirrour or looking-glasse for the ladies of Englande.*

Wherein is deciphered, howe gentlemen under the perfect substance of true love, are oft inveigled with the shadowe of lewde lust: and their firme faith, brought a sleepe by fading fancie: until with joined with wisedome, doth it awake it by the helpe of reason (London: Thomas Dawson for Thomas Woodcocke, 1583).

61. Robert Greene, *Mamilia: The Triumph of Pallas* (London: Printed by Henry Middleton for William Ponsonby, 1583). Thomas Creede printed this title for Henry Middleton again in 1593.

62. Robert Greene, *Gwydonis. The carde of Fancie, Wherein the folly of those Carpet Knights is deciphered, which guyding their course by the compasse of Cupid, either dash their ship against most dangerous rocks, or els attaine the haven with paine* (London: William Ponsonby, 1584); Robert Greene, *The myrrour of modestie: wherein appeareth as in a perfect glasse howe the Lorde delivereth the innocent from all imminent perils, and plagueth the bloudthirstie hypocrites with deserved punishments* (London: Imprinted by Roger Warde, dwelling at the signe of the Talbot neere unto Holburne Conduit, 1584); and Robert Greene, *Morando the tritameron of love: wherein certaine pleasaunt conceites, uttered by divers worthy personages, are perfectly discoursed, and three doubteful questions of love, most pithely and pleasantly discussed* (London: Printed by J. Kingston and J. Charlewood for Edward White, and are to be solde at his shoppe, at the little north deoore of S Paules Church, at the signe of the Gunne, 1584).

63. The following titles by Robert Greene were released between 1585 and 1589: *Planetomachia; or, The first parte of the generall opposition of the seven planets* (London: Printed by Thomas Dawson and G. Robinson for Thomas Cadman, dwelling at the great north doore of Paules, at the signe of the Byble, 1585); *Euphues his censure to Philautus: wherein he is presented a philosophicall combat betweene Hector and Achylles, . . .* (London: Printed by John Wolfe for Edward White, 1587); *Greenes carde of fancie* (London: Printed for William Ponsonby, 1587) [a reprint of *Gwydonis*]; *Morando the tritameron of love: the first and second part* (London: Printed by John Wolfe for Edward White, 1587) [this is an expanded edition of the first printing that represents White's continuing investment in the title]; *Pandosto: the triumph of time. Wherein is discovered by a pleasant historie, that although by the meanes of sinister fortune truth may be concealed, yet by time in spight of fortune it is most manifestlie revealed* (London: by Thomas Orwin for Thomas Cadman, dwelling at the signe of the Bible neere unto the north doore of Paules, 1588); *Perimedes the blacke-smith: a golden methode how to use the minde in pleasant and profitable exercise: wherein is contained speciall principles fit for the highest to imitate, and the meanest to put in practise* (London: Printed by John Wolfe for Edward White, 1588); *Ciceronis amor: Tullies love: Wherein is discoursed the prime of Ciceroes youth, . . .* (London: Printed by Robert Robinson for Thomas Newman and John Winington, 1589); *The Spanish masquerado: wherein under a pleasant devise, is discovered effectuallie, in certaine breefe sentences and mottos, the pride and insolencie of the Spanish estate* (London: Printed by Roger Ward for Thomas Cadman, 1589); *Arbasto: The anatomie of fortune. Wherein is discoursed by a pithie and pleasant discourse, that the highest state of prosperitie is oft times the first steppe to mishappe* (London: Printed by John Charlewood for Henry Jackson, 1589).

64. For three of these titles, John Wolfe printed the books and Edward White sold them. John Wolfe was also an important affiliate of Robert Greene's: he printed eight titles in all, and of the five not sold by White, two were sold by William Wright and three by Thomas Newman, who partnered with Thomas Scarlet and Thomas Gubbin to sell another Greene book.

65. Robert Greene, *A notable discovery of coosenage, now daily practiced by sundry lewd persons called connie-catchers, and crosse-byters* (London: Printed by John Wolfe for T. N. [Thomas Newman], 1591); Robert Greene, *The Second part of conny-catching* (London: Printed by John Wolfe for William Wright, 1591).

66. Robert Greene, *The Third and Last Part of Conny-catching* (London: Printed by T. Scarlet for C. Burby, 1592); Cuthbert Conny-catcher (a.k.a. Robert Greene) *The defence of conny catching* (London: Printed by A. Jeffes for Thomas Gubbins to be sold by John Busbie, 1592); *A disputation between a hee conny-catcher and a shee conny-catcher: whether a theefe or a whoore is most hurtfull in cousonage, to the common-wealth. Discovering the secret villainies of alluring strumpets* (London: A. Jeffes for T. Gubbins, 1592).

67. Robert Greene, *The blacke bookes messenger: Laying open the life and death of Ned Browne one of the most notable cutpurses, crosbiters, and conny-catchers, that ever lived in England* (London: Printed by John Danter for Thomas Nelson, 1592).

68. *The second part of conney-catching* was printed by John Wolfe for William Wright again in 1592 and 1597. See, for example, *The groundworke of conny-catching, the maner of their peddlers-French, and the meanes to understand the same* (London: n.d. [not before 1592]).

69. Joseph Loewenstein, "The Script and the Marketplace," *Representations* (1985) 12: 101–14. For a broader view of authors and the context of writing popular works during this period, see Peter Saunders Webb, *Writing in the Bearpits: Popular Authors in Early Modern England* (Ph.D. diss., University of Michigan, 1991), esp. ch. 1.

70. It is largely agreed that Greene wrote this work, and it seems likely that the author, the printer, or the pair were playing with his academic history to hint to readers the true nature of the book. Even if the title claimed that it was a refutation of the first two coney-catching books, the pseudonym promised that it would be a similar walk through the criminal population. It also poked fun at the academic tendency toward long-winded, high-minded refutations of each other's work, a tradition that every browser would recognize, since those books appeared on the same shelves as less insular works.

71. The full title of *Greenes mourning garment,* for example, reads *Greenes mourning garment: given him by repentance at the funerals of love, which he presents for a favour to all young gentlemen that wish to weane themselves from wanton desires,* R. Greene Utriusque academiae in Artibus magister (London: Printed by J. Wolfe for T. Newman, 1590); the second title is no more subtle: *The Repentance of Robert Greene, Maister of Arts. Wherein by himselfe is laid open his loose life with the manner of his death* (London: Printed by J. Danter for Cuthbert Burbie, 1592).

72. The pamphlet war and attacks on Greene's character by Gabriel Harvey and his followers after his death would have helped to keep him in the public imagina-

tion and enforce his reputation as a scoundrel. For a review of these attacks and the responses that his friends and followers wrote, see Alexander B. Grosart, ed., *The Life and Complete Works in Prose and Verse of Robert Greene, MA* (London: for private circulation only, 1881–86), 1: 1–53.

73. *Orlando Furioso* was printed by J. Danter for Cuthbert Burbie in 1594 and *The honorable historie of friar Bacon and friar Bongay* was printed by Adam Islip for Edward White the same year.

74. *Morando the tritameron of love wherein certaine pleasant conceits, uttered by divers worthy personages, are perfectly discoursed, and three questions of love, most pithily and pleasantly discussed: shewing to the wyse how to use love, and to the fonde, howe to eschew lust: and ye pleasure and profit* (London: Printed by John Kingston and John Charlewood for Edward White, 1584).

75. *Euphues his censure to Philautus wherein is presented a philosophicall combat betweene Hector and Achylles, discovering in foure discourses, interlaced with diverse delightfull tragedies, the vertues necessary to be incident in every gentlemen* by Robert Greene, in *Artibus Magister* (London: Printed by John Wolfe for Edward White, 1587).

76. *Perimedes the Blacke-smith, A golden methode, how to use the minde in pleasant and profitable exercise: Wherein is contained speciall principles fit for the highest to imitate and the meanest to put into practise, how best to spend the wearie winter nights, or the longest summers Evenings, in honest and delightfull recreation. Wherein we may learne to avoide idlenesse and wanton scurrilitie, which divers appoint as the end of their pastimes. Herein are interlaced three merrie and necessarie discourses fit for our time: with certaine pleasant Histories and tragical tales which may breed delight to all, and offence to none, Omne tulit punctum, qui miscuit utile dulci* (London: Printed by John Wolfe for Edward White, 1588).

77. *Philomela, the Lady Fitzwaters Nightingale, By Robert Greene, Utriusque Academiae* in *Artibus magister*, serosed serio (London: Printed by Robert Bourne for Edward White, 1592).

78. David Hall, *Worlds of Wonder, Days of Judgment: Popular Religious Belief in Early New England* (Cambridge, Mass.: Harvard University Press, 1990).

Chapter 2: Roger Bacon, Robert Greene's Friar Bacon, and the Secrets of Art and Nature

1. Robert Greene, *The honorable historie of frier Bacon and frier Bongay* (Printed for Edward White: sold at his shop at the little North door of Paules at the signe of the gun, 1594), B2, v.

2. Helgerson, *Forms of Nationhood*, ch. 1.

3. Eamon, *Science and the Secrets of Nature*, 55.

4. Fissell, "Hairy Women and Naked Truths"; Roy Porter, "'The Secrets of Generation Display'd': *Aristotle's Masterpiece* in Eighteenth-Century England," *Eighteenth Century Life* 9,3 (1985): 1–21.

5. Steven Shapin, *A Social History of Truth: Civility and Science in Seventeenth-Century England* (Chicago: University of Chicago Press, 1994).

6. Mary Fissell, "Readers, Texts, and Contexts: Vernacular Medical Works in Early Modern England," in *The Popularization of Medicine, 1650–1850*, edited by Roy Porter (London: Routledge, 1992), 72–96.

7. Brian Copenhaver, "Natural Magic, Hermetism, and Occultism in Early Modern Science," in *Reappraisals of the Scientific Revolution*, edited by David C. Lindberg and Robert S. Westman (Cambridge, U.K.: Cambridge University Press, 1990), 261–301.

8. David C. Lindberg, *Roger Bacon's Philosophy of Nature: A Critical Edition, with English Translation, Introduction, and Notes, of De multiplicatione specierum and De speculis comburentibus* (Oxford, U.K.: Clarendon Press, 1983), xv–xxvi.

9. David Lindberg (*Roger Bacon's Philosophy*, xv) argues that the translation of the *Secretum Secretorum* was not a crucial moment for Bacon, who moved away from Aristotle and toward contemporary writers after this translation. William Eamon, on the other hand (*Science and the Secrets of Nature*, 45–53), argues that this text was important for Bacon and other medieval and early modern occult practitioners.

10. Lindberg, *Roger Bacon's Philosophy*, xviii–xix.

11. A. C. Crombie and J. D. North, "Bacon, Roger," *Dictionary of Scientific Biography* (Oxford, U.K.: Oxford University Press, 1970), 1: 377–85.

12. The family supported Henry III against Simon de Montfort and his barons, resulting in their financial downfall. Lindberg, xvi; Roger Bacon, *Opera quaedam hactenus inedita*, edited by J. S. Brewer (London: n.p., 1859), 16.

13. Stewart Easton, *Roger Bacon and His Search for a Universal Science* (Oxford, U.K.: Oxford University Press, 1952), 118–26.

14. Lindberg, *Roger Bacon's Philosophy*, xx.

15. Ibid., xxiii.

16. Ibid., xxiii–xxiv.

17. The nuances of this story lie far outside the scope of this chapter, but Lindberg has amassed good evidence for the presence of these texts at the papal court by tracing their influence on the work of the Polish mathematician Witelo. David C. Lindberg, "Lines of Influence in Thirteenth Century Optics: Bacon, Witelo, and Pecham," *Speculum* 46 (1971): 72–75.

18. Lindberg, *Roger Bacon's Philosophy*, xxv–xxvi.

19. John Dee's name and pursuits were firmly ensconced in the Elizabethan court, which suggests that Bacon's name and works also had some currency there. For the textual link between Dee and Bacon, see Roberts and Watson, *John Dee's Library Catalogue*.

20. McKerrow notes that Edward White was a London bookseller under the sign of the gun and at the little north door of St. Paul's between 1557 and 1612. He was apprenticed to William Lobley between 1565 and 1572, with his first entry in the registers in 1576. He dealt primarily in ballads and was fined 10 shillings in June 1600 for printing one called *The Wife of Bath* (presumably not Chaucer's). His business was continued after his death by his wife Sarah, and his son followed him into the printing trade.

R. B. McKerrow, ed., *A Dictionary of Printers and Booksellers in England, Scotland, and Ireland, and of Foreign Printers of English Books 1557–1640* (London: Printed for the Bibliographical Society by Blades, East, and Blades, 1910), 289. For a record of early performances of this play, see R. A. Foakes and R. T. Rickert, *Henslowe's Diary* (Cambridge, U.K.: Cambridge University Press, 1961), 16–21, and James A. Levin, ed., *Friar Bacon and Friar Bungay* (London: Ernest Benn, 1969), xiii.

21. Foakes and Rickert, *Henslowe's Diary*, 13–17.

22. The decision to name the printed edition a "good quarto" was made most clearly by Scott McMillin, "The Queen's Men in 1594: A Study of 'Good' and 'Bad' Quartos," *English Literary Renaissance* 14,1 (1984): 55–69.

23. Kerstin Assarsson-Rizzi, *Friar Bacon and Friar Bungay A Structural and Thematic Analysis of Robert Greene's Play* (Berlingska: Lund, 1972), 28; Charles Hieatt, "A New Source for Friar Bacon and Friar Bungay," *Review of English Studies* 32,126 (1981): 180–87, see esp. 180.

24. Grosart, xxx–xxxi.

25. He was a prolific writer of pamphlets and plays, producing consistently from his 1579 pamphlet *Mamilia* (printed in 1593) through his deathbed *Repentance* in 1592. He was attacked for his wanton lifestyle and decried as a poor and unoriginal author by Gabriel Harvey and his followers almost immediately after his death, but he was also defended and praised by his contemporaries, including his friends and the anonymous R. B., who wrote "Greene's Funerals" in 1594 (Grosart, 1: 1–53).)

26. Nearly everyone who has written on this play since 1881 agrees that Greene borrowed some plotline and structure from Lyly's *Campaspe*; the idea appears in Grosart, who reports it as an accusation hurled by Harvey after Greene's death, and in modern criticism in Hieatt, 180–87. The tie to Marlowe has been debated, since neither work has a certain date of creation. For a strong assertion of the inevitable inconclusiveness of this argument, see Assarsson-Rizzi, 14–15, 148.

27. Arber, *Transcript of the Registers of the Company of Stationers*, vols. 1–3.

28. *The Famous historie of Frier Bacon* first appears in the register in 1624. While it is possible that it circulated as a manuscript, the audience it could have reached in that form is much smaller than the one it reached when printed legally and inexpensively for 60 years. It is also possible that an entirely undiscovered and unrecorded printing was made of this text, but the evidence currently available suggests that it did not appear until the seventeenth century.

29. Paul Dean, "Shakespeare's Henry VI Trilogy and Elizabethan 'Romance' Histories: The Origins of a Genre," *Shakespeare Quarterly* 33,1 (1982): 34–48.

30. Douglas L. Petersen, "Lyly, Greene, and Shakespeare and the Recreations of Princes," *Shakespeare Studies* 20 (1987): 67–88.

31. I strongly disagree with Tetzeli von Rosador's reading of the depiction of magic and its relationship to the state in this play ("The Sacralizing Sign: Religion and Magic in Bale, Greene, and Early Shakespeare," *Yearbook of English Studies* 23 [1993]: 30–45, esp. 38–40). I would argue that, rather than the state "delimiting, containing, or annihilating" magic, it claims and makes use of it to extend national boundaries and influence, and in so doing, it lends legitimacy to occult or magical practices. This

is vastly different from Tetzeli von Rosador's argument, which ends with magic having only a very small and nearly illegitimate place in the play. He accomplishes his argument by ignoring one of the most significant of Bacon's scenes, in which the friar competes with Vandermast to see which nation (England or Germany) has produced the better occult practitioner. Bacon wins in a remarkable display of skill, bringing glory to himself and England, demonstrating the value of his talent to the court and underlining that value through the conjuring of a tremendous feast gathered from around the world. In this case, it is Bacon's skill that gives England imperial scope.

32. Foakes and Rickert, *Henslowe's Diary*, 172.

33. Friar Bongay is a talented magus but he has not yet bent the natural and preternatural worlds to his will. He is, however, happy to meddle in human affairs, and his meddling drives the plot when he marries Lacy and Elinor. Some scholars have said that Bongay is based on Albertus Magnus. I cannot imagine why Greene would have used Bacon's name and not Albert's, but neither can I find evidence in the play to definitively prove that Bongay is not based on Albert. My conjecture is that the resemblance, if intentional, is slight and that Bongay operates in a supporting role that highlights Bacon's mastery, rather than making a specific comparison between the achievements of Bacon and Albert.

34. Peter Stallybrass discusses the importance of clothing and its exchange, particularly in relationship to the stage, in his essay "Worn Worlds" (in *Subject and Object in Renaissance Culture,* edited by Margreta De Grazia, Maureen Quilligan, and Peter Stallybrass [Cambridge, U.K.: Cambridge University Press, 1996], 289-320. He notes that characters switching costumes and, with them, personality traits, was a common theme in Renaissance plays, which were, furthermore, performed in a context that related clothing, personal identity, and interpersonal alliances.

35. *Frier Bacon and frier Bongay,* A4.

36. This will be discussed at length in the next chapter. The example is taken from the anonymous *Cornucopiae.*

37. *Frier Bacon and frier Bongay,* A4.

38. Ibid., B1.

39. Ibid.

40. Stanton Linden (*The Mirror of Alchimy by the Thrice-Famous Friar Bacon* [New York: Garland Press, 1992], xxxiii-xxxviii) cites both these quotations in his discussion of Greene's play that accompanies his edition of Bacon's alchemical text, *The Mirror of Alchimy.* His reading of these lines and of the entire play, however, is very straight—he largely overlooks the multiple depictions of Bacon and the multiple transitions in Bacon's relationship to occult knowledge.

41. *Frier Bacon and frier Bungay,* B2.

42. Frances Yates notes the increasing power of the accusations of conjuring leveled against John Dee after his return to England in 1589. She argues that the loss of court support rendered him and his practices defenseless against his increasingly vocal critics, resulting in a letter to the archbishop of Canterbury (printed in 1604 but written earlier) in which Dee claimed that he pursued his work with divine reverence and with the hope of understanding divine truth. "The cry of 'conjurer' had always been spo-

radically raised but in the old days the queen and Leicester had protected his studies" (*The Occult Philosophy in the Elizabethan Age* [London: Routledge, 1979], 105).

43. *Frier Bacon and frier Bungay*, B2.

44. Ibid. Lynn Veach Sadler ("Alchemy and Greene's *Friar Bacon and Friar Bungay*," *Ambix* 22,2 [1975]: 113, 115) notes that the real Bacon believed that the most powerful adepts combined alchemy and experimental natural philosophy in an esoteric pursuit that could benefit the state.

45. *Frier Bacon and frier Bongay*, B3.

46. Ibid.

47. Ibid.

48. Ibid., A4.

49. The metaphor of the chemical wedding has been explored in great detail elsewhere. For a clear explanation, see Charles Nicholl, *The Alchemical Theater* (London: Routledge, 1980), 38–39. For a fascinating study on a female alchemist's appropriation and revision of the metaphors of generation in the art, see Tara E. Nummedal, "Alchemical Reproduction and the Career of Anna Maria Zieglerin," *Ambix* 48,2 (2001): 56–68.

50. Allison Kavey, "Mercury Falling: Gender Flexibility and Eroticism in Popular Alchemy," in *The Sciences of Homosexuality in Early Modern Europe*, edited by George Rousseau and Kenneth Borris (New York: Routledge, 1998).

51. *Frier Bacon and frier Bongay*, B4.

52. Ibid.

53. Ibid., F1.

54. Ibid.

55. Ibid., F3.

56. Ibid., G3.

57. Ibid., C3.

58. Ibid., H5.

59. Ibid., I2.

60. Thomas Creede was among the network of printers of books of secrets and other cheap texts. He is particularly notable in this case for also having printed one of Greene's plays, *James IV*.

61. *The Mirror of Alchimy* has already been productively analyzed by several scholars, including Stanton J. Linden, who produced an edition of the text in 1992: *The Mirror of Alchimy: Composed by the Thrice-Famous and learned fryer, Roger Bacon* (New York: Garland Books, 1992); and Charles Nicholl, *The Chemical Theater* (London: Routledge, 1980), ch. 2.

62. *Mirror of Alchimy*, title page.

63. Yates, *Occult Philosophy*.

64. *Trismegistus* translates as "thrice-greatest." Brian Copenhaver has devoted a great deal of attention to tracing the story behind the Emerald Table. He points out that it was not, in fact, an ancient Egyptian text but one created by a medieval scholar and circulated under false pretenses throughout the late medieval and early modern period: "Natural Magic."

65. *Mirror of Alchimy*, 17.

66. Ibid., 18.

67. Hortulanus, "A brief Commentarie of Hortulanus," in *Mirror of Alchimy*, 17–27, esp. 18.

68. Galid, "The Booke of the Secrets of Alchimie," in *Mirror of Alchimy*, 28–53, esp. 28.

69. Roger Bacon, "The Admirable Force of Art and Nature," in *Mirror of Alchimy*, 54–84, esp. 77.

70. Edmund Brehm, "Roger Bacon's Place in the History of Alchemy," *Ambix* 23,1 (1976): 53–58.

71. Ibid., 54–84, esp. 74. Emphasis in original.

72. For an absolutely remarkable dissection of the coding in Boyle's alchemical writing, see Lawrence M. Principe, "Robert Boyle's Alchemical Secrecy: Codes, Ciphers, and Concealments," *Ambix* 39,2 (1992): 63–74. Edmund Brehm ("Bacon's Place," 54) seems amazed at Bacon's vagary in his alchemical writing and believes that it suggests a lack of practical alchemical knowledge. His interpretation may be indebted to his approach, which required that he examine Bacon's writings for precursors to modern chemical thinking and techniques. This angle requires an analysis of these works that removes them from their place in the medieval occult and natural philosophical tradition, one known for its coding and secrecy, not for its clear communication of techniques.

73. *Mirror of Alchimy*, 1, 76.

74. The precise description of the art notory reads "an art whereby a man may write or note anything, as briefly as he will, and as swiftly as he can desire" *Mirror of Alchimy*, 78. The study of deliberate obfuscation and the use of codes in alchemical writing has already been pursued by Lawrence Principe ("Boyle's Alchemical Secrecy"). He records some of the same methods advertised in this treatise, including the use of foreign letters and languages, ciphers, and a form of shorthand.

75. Bacon, "The Admirable Force of Art and Nature," in *Mirror of Alchimy*, 54–84, esp. 78.

76. Ibid., esp. 80.

77. *Mirror of Alchimy*, esp. 7.

78. Ibid., esp. 8.

79. This theory, when applied to common people, is derived from the use of catechism to teach children to read, as well as the assumption that most readers had access to only a few books and thus reread them frequently. Elite readers also appear to have read intensively, making repeated attempts to understand the texts inherited from antiquity. On this point, see Rolf Engelsing, *Der Burger als Leser, Lesergeschichte in Deutschland, 1500–1800* (Stuttgart, Ger.: Metzlersche, 1974).

80. Hortulanus, "A briefe Commentarie," 17–27, esp. 18.

81. Ibid., 29.

82. Ibid., 30.

83. These stages are featured in both "The Myrrour of Alchimy" and "The Booke of the Secrets of Alchimie," both in *Mirror of Alchimy*, but the first text (pp. 1–16) of-

fers a great deal more description of the processes, while the second (28–53) primarily lists the colors through which it passes. See *Mirror of Alchimy,* 12–14, 48.

84. Galid, 50.
85. Bacon, "An excellent discourse of the admirable force and efficiencie of Art and Nature," *Mirror of Alchimy,* 57.
86. Ibid., 55.
87. Ibid., 76.
88. Bacon, "Myrrour of Alchimy," 9.
89. Ibid.
90. Ibid., 11.
91. Galid, 32.
92. One thing that remains to be considered is what might happen to the believability of this model of the natural world if the arts designed to mimic it never worked. I think it would probably remain unchanged, with an understanding that the failure lay in the hands of the alchemist who had failed to properly follow the steps, rather than in the larger understanding of the world.
93. Galid, 36.
94. Ibid., 40.
95. Ibid., 48–52.

Chapter 3: Structuring Secrets for Sale

1. Copeland 1565 edition.
2. *Cornucopiae.*
3. *The Secrets of Alexis of Piedmont* and *Certen secrets of nature,* both of which were published during this period and contain presentations of natural knowledge similar to those in the books chosen for this chapter, were well out of the price range of most readers because of their length. The excerpts from Pliny's *Secrets and wonders of the world* would have been affordable, but the organization of the English edition is highly dependent upon the original text.
4. The four printings were all done in London, by J. King, 1560; William Copeland, 1560 and 1565; William Seres, circa 1570; and William Jaggard, 1599.
5. This theory asserted that divine planning had made it so that every object was connected to other objects in the natural world through shared characteristics. Furthermore, these relationships were accessible to trained and observant people who could use them to "read" the book of nature. It played an important role in medieval and Renaissance natural philosophy and medicine and played a central role in Paracelsus's model of the world. For more on systems of sympathy, see Paula Findlen, "Empty Signs? The Book of Nature in Renaissance Science," *Studies in History and Philosophy of Science* 21,3 (1990): 511–18.
6. Roger Chartier, "Popular Appropriation: Readers and Their Books," *Forms and Meanings: Texts, Performances, and Audiences from Codex to Computer,* edited by Roger Chartier (Philadelphia: University of Pennsylvania Press, 1995).
7. Ibid., 89.

8. Roland Barthes, *S/Z*, translated by Richard Miller (New York: Hill & Wang, 1970).

9. "They [secrets] included, in the first place, manifestations of occult qualities, or events that occur unexpectedly or idiosyncratically as a result of occult causes.... Secreta also referred to certain events that took place as a result of artificial instead of natural causes, tricks of the trade as it were.... Secreta could be experienced, but because they were not demonstrable, they could not be the objects of scientific knowledge." Eamon, *Science and the Secrets of Nature*, 54.

10. Fox, *Oral and Literate Culture*.

11. Barthes, *S/Z*.

12. Unfortunately, very little is known about the details of this job. The best evidence for the relationship between compilers and printers can be found in letters to readers, some of which do seem to have been the creation of the book's compiler or translator. In the case of *The Secrets of Albertus Magnus*, the first three English editions share the same letter, which may have been the work of the anonymous translator. The last edition has a markedly different letter, which makes it difficult to generalize about the translator's impact even across editions of the same book. The translator of *Cornucopiae*, Thomas Johnson, has his name on the frontispiece of both editions, but he did not write a letter, or if he did, it is not included in either edition.

13. King 1560 edition, Aii^r; Copeland 1565 edition, Aii^v.

14. Jaggard 1599 edition, Aiii^v.

15. Jaggard 1599 edition, Aiii^v.

16. Jaggard 1599 edition, Aii^{r-v}.

17. Jaggard 1599 edition, Dviii^r.

18. Two other categories attracted enough attention to warrant two entries each: overcoming enemies and seeing the future.

19. Jaggard 1599 edition, Ci^r, Dv^r, Dviii^r.

20. "Moreover, if this stone be put, braed, and scattered upon coles, in foure corners of the house, they that be sleapinge, shall flee the house, and leave all." Jaggard 1599 edition, Ci^r.

21. Jaggard 1599 edition, Dvi^r.

22. Jaggard 1599 edition, Dviii^v.

23. Jaggard 1599 edition, Fii^v.

24. Jaggard 1599 edition, Fii^v–Fiii^v.

25. Jaggard 1599 edition, Fiv^r.

26. Jaggard 1599 edition, Hi^v.

27. Jaggard 1599 edition, Hiii^v.

28. The Latin in this quotation is remarkably tortured, and I owe many thanks to Larry Principe and Steven vanden Broecke for contributing to my efforts to decode it. "She returns to being upright upon herself, as if she had attained from the male the virtue of the male by means of fumes through joining together." Jaggard 1599 edition, Hiii^r.

29. Jaggard 1599 edition, Hiii^r.

30. "The first herbe is called with the men of Chaldea Elios, with the Grekes Matu-

chiol, with the Latines Elitropium, with Englyshe men Marygolde, whose interpretation is at Elion, that is the Sunne, and Tropos, that is alteration or change, because it is turned according to the sun." Copeland 1565 edition, Aiii^v.

31. Copeland 1565 edition, Aiv^v.
32. Copeland 1565 edition, Jiii^r.
33. Copeland 1565 edition, Av^v.
34. Copeland 1565 edition, Evi^r.
35. "It returns forgetfulness and loss of reason to him who remembers." Copeland 1565 edition, Ji^v and Jiiii^v.
36. *Cornucopiae*, A3^r.
37. *Cornucopiae*, A3^r.
38. *Cornucopiae*, A4^r.
39. *Cornucopiae*, C2^r-C4^v.
40. For more on this trend in natural philosophy, see Shapin, *Social History of Truth*.
41. Copeland 1565 edition, Fviii^r.
42. "Put thou this herb with the heart of a young frog, and her matrix, and put them where thou will, and after a little time all the dogs of the whole town shall be gathered together. And if thou shall have the aforenamed herb under they foremost toe, all the dogs shall keep silence, and have no power to bark. If thou put the aforesaid thing in the neck of any dog (so that he may not touch it with his mouth) he shall be turned aways round about like a turning wheel, until he fall unto the ground as dead, and this hath been proved in our time." Copeland 1565 edition, Avi^v.
43. I am referencing the effect of pennyroyal, bay, or laurel leaves, and the skin of the black plover discussed above, Copeland 1565 edition, Avi^v-Avii^r; the Avicenna attribution can be found on Fviii^r.
44. Copeland 1565 edition, Fviii^v-Gi^v.
45. Copeland 1565 edition, Gii^v.
46. Copeland 1565 edition, Biii^r, Dviii^r, and Fii^r.
47. Copeland 1565 edition, Fviii^v.
48. Copeland 1565 edition, G1^r.
49. Shapin (*Social History of Truth*, ch. 2 and 3) has established the importance of witnessing in the context of aristocratic and especially Royal Society natural philosophy. The remarkable thing about these two books is that they emphasize witnessing a century before the subject of Shapin's work, and they privilege the witnessing abilities of a population of readers from nonaristocratic backgrounds who had little formal education. In other words, they privilege the very readers that Shapin discounts as not having the social position to act as reliable witnesses in early modern England.
50. *Cornucopiae*, A3^v.
51. Copeland 1565 edition, Av^{v-r}.
52. Copeland 1565 edition, Dvi^{v-r}.
53. Copeland 1565 edition, Gii^r and Ii^r.
54. Examples of medieval attempts to explain the magnet's operation in natural terms can be found in J. L. Heilbron, *Electricity in the Seventeenth and Eighteenth*

Centuries: A Study of Early Modern Physics (Berkeley: University of California Press, 1979).

55. *Cornucopiae*, A3^v.
56. *Cornucopiae*, A4^v.
57. *Cornucopiae*, A5^r.
58. *Cornucopiae*, B2^v.
59. *Cornucopiae*, A4^v-B1^r.
60. *Cornucopiae*, F3^r.
61. *Cornucopiae*, B3^v.
62. *Cornucopiae*, F2^v.

Chapter 4: Secrets Gendered

1. John Partridge was a successful poet and translator in sixteenth-century London. He translated a number of history plays with strong national themes and headed them with poems celebrating the glory of England and the importance of a strong, well-intentioned and well-advised monarch for its continuation and expansion. For information on Richard Jones, see chapter 1.

2. These letters remain in later editions of this book, even those produced by other printers. See, for example, the Henry Car edition of 1586 (British Library Humanities Collection, C31.a15). Car's is an interesting edition for other reasons as well, since it reflects some of the new recipes that appeared in Jones 1596 *Treasurie of hidden Secrets* and some recipes that appear in neither of the Jones editions nor the Edward White copies of *The Good Huswifes Jewell, The Seconde Parte of the Good Huswives Jewell*, nor *The Good Huswives Handmaid for Cookerie in her Kitchin*.

3. There is no record of a Richard Jones edition of this book, but Edward White printed a book around 1595 that may be the text in question: *The Good Huswives Handmaid for Cookerie in her Kitchin* is a collection of recipes, primarily for meats, poultry, and fish, that is particularly interesting because it also contains instructions for the order of courses at a banquet and a recipe for making dishes out of sugar, a staple of elite dinner parties that epitomized the place of trifles in Elizabethan culture. It has, however, very few medical recipes. It is packaged with White's 1595 edition of *The Widowes Treasure*, the 1596 edition of *The Good Huswifes Jewell*, and 1597 edition of *The Second Part of the Good Huswifes Jewell*.

4. This strategy is a combination of literary analytic techniques employed by Nancy Armstrong and Leonard Tennenhouse (*The Ideology of Conduct: Essays on Literature and the History of Sexuality* [New York: Methuen, 1987]) in their considerations of power and gender in eighteenth-century imperial fiction and Eve Kosofsky Sedgewick's (*Epistemology of the Closet* [Berkeley: University of California Press, 1991]) framework for thinking about the function of homoerotics and homosexuality for structuring intellectual and social worlds in nineteenth-century fiction. While these techniques were developed for later and intentionally "fictional" texts, they are no less useful for sixteenth-century books of secrets, which were, after all, simply products of their own cultural moment, a moment that paid less heed to the currently consuming

division between truth and make-believe or, in this case, science, medicine, cooking, and magic.

5. Crawford and Laura, *Women's Worlds*; Mendelson and Crawford, *Women in Early Modern England*—esp. Introduction; Gowing, *Domestic Dangers*; Crawford, *Women and Religion in England,*; Mendelson, *Mental World of Stuart Women.*

6. For examples of this kind of work, see Paula Findlen, "Science As a Career in Enlightenment Italy: The Strategies of Laura Bassi," *Isis* 84,3 (1993): 440–69; Pelling, "The Women of the Family?"; Pennell, "Pots and Pans History"; Stine, "Opening Closets."

7. David Cressy (*Travesties and Transgressions in Early Modern England* [Oxford, U.K.: Oxford University Press, 2000]) notes that predictable natural occurrences reassured early modern people that their world was functioning normally, while breaks in that pattern indicated a disturbance of the natural or supernatural order. The importance of recipes, which are assumed to produce the same results each time they are followed, becomes clearer in this interpretation—a failed recipe could mean not just a bad batch of conserve or a scatterbrained housewife, but a shift in the natural order. David Underdown, "The Taming of the Scold," in *Order and Disorder in Early Modern England,* edited by Anthony Fletcher and John Stevenson (Cambridge, U.K.: Cambridge University Press, 1987); Gowing, *Domestic Dangers.*

8. Partridge, *Treasurie of commodious Conceits,* Aiii^v.

9. Recipe books, both in manuscript and in print, are a burgeoning topic in print, women's, and cultural history. An example, cited earlier, is Pennell, "Pots and Pans History."

10. This seems to follow the model of the inclusion of exotic ingredients in medicines. While many women probably lacked the gold, pearls, or other expensive ingredients in medical recipes, their inclusion in popular medical manuals contributed to medical practitioners' thinking of healing outside a local and herbal context.

11. Partridge, *Treasurie of commodious Conceits,* Ei.

12. David Cressy (*Travesties and Transgressions,* 37) notes that "The cycle of birth and death, like the cycle of the seasons, was mostly normal, conformable, and predictable, though subject to occasional surprises, freaks, and quirks. Regular patterns were a sign of good order, all right with the world."

13. Jones 1596 *Treasurie of hidden Secrets,* Aii^r.

14. Mendelson and Crawford, *Women in Early Modern England,* 217.

15. Two essays in Gordon McMullan's *Renaissance Configurations—Voices/Bodies/Spaces* laid the groundwork for my thinking about closets and secrets. James Knowles ("'Infinite Riches in a Little Room': Marlowe and the Aesthetics of the Closet") writes extensively on the differences in use and meaning of men's and women's closets in early modern England. Sasha Roberts ("Shakespeare 'Creepes into the Women's Closets about Bedtime': Women Reading in a Room of Their Own") writes instructively about the social and architectural place of the closet in the sixteenth-century house, gender differences in the use and meaning of closets, and the importance of locks.

16. Jones 1596 *Treasurie of hidden Secrets.*

17. The strategy of attributing the decision to print to an anonymous person or

persons, particularly those with good social standing, was very common in the world of early modern cheap print. The gentlewoman in this case is interesting because of her sex and because of the strange definition of secrecy she encourages Partridge to adopt in order to justify printing his book.

18. Partridge, *Treasurie of commodious Conceits*, Aiii^r.

19. Mendelson and Crawford reflect on the extensiveness of female conversational networks and the importance of gossip in portraying and regulating behavior within communities and passing useful domestic, historical, literary, and mythical information between the generations and the sexes. "Speech was the primary medium for transmitting not only superstitions and magical lore, but collective feminine experience about housewifery, medicine (particularly gynecology and obstetrics), gardening, cookery, childcare, textile, and other work skills, and a host of other philosophical as well as practical concerns" (*Women in Early Modern England*, 216–18).

20. Books like these seem to fit into Fox's space between reading out loud and silent reading, since the design of the book allows for each recipe to be shared, considered, and discussed.

21. *Widowes Treasure*, Aii^r.

22. Mendelson and Crawford note that widows could be very powerful women if their husbands had managed their affairs well and they either had no sons or had sons who allowed them to keep and control their own property. They maintained the social status they had held as wives and were more likely to attract sympathy and instrumental support than other adult women living without men. This was not always the case, however, and many women accused of witchcraft were also widows who were suspected of lustfulness and mischief because they were no longer under the direct control of men. The fact that they were also frequently required to work for their survival further jeopardized their social status, because the positions available to women tended to rely on their knowledge of female secrets such as childbirth and herbal medicine, which could raise the ire of suspicious men. See Stuart Clark, *Thinking with Demons: The Idea of Witchcraft in Early Modern Europe* (Oxford, U.K.: Oxford University Press, 1999).

23. *Widowes Treasure*, Aii^v.

24. Ibid. Aiii^v.

25. Ibid.

26. Ibid.

27. Jones 1596 *Treasurie of hidden Secrets*, frontispiece.

28. D. A. Miller, *The Novel and the Police* (Berkeley: University of California Press, 1998).

29. Jones 1596 *Treasurie of hidden Secrets*, Aiii^v.

30. Ibid.

31. Ibid.

32. Partridge, *Treasurie of commodious Conceits*, Cii^{v-r}.

33. Ibid., Cii^v–Cv^v.

34. Ibid., Dii^{v-r}.

35. Ibid., Eiii^r.

36. Ibid., Fvi^v.
37. Ibid., Fix^v.
38. Jones 1596 *Treasurie of hidden Secrets*, 34–37.
39. Ibid., 45, 46, 48.
40. Ibid., 23, 27–33.
41. *Widowes Treasure*, Aiiii^v.
42. Ibid., Aiiii^r.
43. Ibid., for jellies and fruits, Cii^v–Civ^v; for conserves, Cvi^r; for meats, eggs, and vegetables, Fv^v–Gi^v.
44. Ibid., Avi^r–Aviii^r.
45. Partridge, *Treasurie of commodious Conceits*. Bvi^v.
46. Ibid., Cii^v.
47. Ibid., Dviii^v, Evi^r; Cv^r.
48. Ibid., Evii^v, Fii^r; Fvi^v–Fviii^v.
49. Ibid.: Dii^r, "A receipt to restore strength"; Diii^r, "A violet powder for woolen clothes and furs"; Div^v, "A sweet powder for Napery and all linen clothes"; Dv^r, "To make a pomeamber"; Ei^v, "A powder wherewith to make sweet waters"; Ei^r, "another manner of making Damask water"; Eii^v, "To perfume gloves"; Eiii^v, "A preparative for gloves"; Eiii^v, Eiii^r, and Eiv^v, "another for gloves"; and Ev^v, "Another manner of making Damask water."
50. Ibid., Fii^r.
51. There are four recipes for ink in this text, all on the same page and described according to the richness of their color: "a perfect blacke inke," "a very good Greene," "Another Greene," and "An Emerald greene" (*Widowes Treasure*, Bv^{v-r}).
52. Ibid., Fvi^r.
53. Ibid., Av^v.
54. Ibid., Bi^{v-r}.
55. Ibid., Gi^{v-r}.
56. Jones 1596 *Treasurie of hidden Secrets*: women's headings, 34; powder, 19.
57. Ibid., 35.
58. Ibid., 67.
59. Ibid., 23.
60. Ibid., 23-24.
61. Ibid., 27.
62. Ibid., 28.
63. Ibid., 32.
64. Ibid.
65. Ibid., 33.
66. Ibid.
67. Ibid., 45, 46, 48.
68. Ibid., 46.
69. Ibid.
70. Ibid.

71. Alexandra Walsham, *Providence in Early Modern England* (Oxford, U.K.: Oxford University Press, 1999).

72. Jones 1596 *Treasurie of hidden Secrets,* 48–49.

Chapter 5: Secrets Bridled, Gentlemen Trained

1. Markham, *How to chuse.*

2. Joan Thirsk, "Horses in Early Modern England: For Service, for Pleasure, for Power," in *The Rural Economy of England,* edited by Joan Thirsk (London: Hambledon Press, 1984), 375–401.

3. Ibid., 385.

4. Ibid., 383.

5. Ibid., 380.

6. The details of the rest of the 1530s and 1540s regulations can be found in Thirsk, 383, 384.

7. Robert Markham was Gervase Markham's father and one of Henry's gentlemen pensioners. The family's extensive land holdings and equestrian ties to the court make it hardly surprising that Gervase made his living writing about the breeding and training of horses, as well as other gentlemen's pursuits. The information about the queen's breeding program, the gentry's adoption of her standards, and the practice of gifting horses can be found in Thirsk, 386–88.

8. Thirsk, 388–89.

9. The first printed book on horsemanship appeared in Spain in 1495. It was followed by three more printed in Italy in 1499, 1517, and 1518. By the 1550s, France, Spain, and Germany were producing a plethora of books on the art of training horses for pleasure (Ibid., 389).

10. Ibid., 389–91.

11. Ibid., 389.

12. Ibid., 375–76.

13. Thirsk, "Farming in Kesteven," in Thirsk, *Rural Economy of England,* 137.

14. The highest price recorded for a single animal was 3 pounds, 6s, 8d. This was on the high end of equine prices, although not outside the market rate. Thirsk cites the will of Thomas Wright, a farmer who died still owing money for livestock, including 3£, 6s, 8d for a mare and foal, 3£, 9s, 2d for a horse, and 33s, 4d for a bag nag (Thirsk, "Farming," 139; "Horses," 393–94).

15. Thirsk, "Horses," 397.

16. The importance of oral communication in early modern literacy has certainly not been entirely ignored. Some excellent current work on this difficult problem can be found in Fox.

17. The section on building a good horse farm gives evidence of the financial status of the intended readers. The land required to do this kind of building was expensive and was usually inherited rather than purchased. The number of people required to run such an operation is extensive, and thus expensive—servants required board and

wages. Markham's insistence on using an Arabian stallion, which would have had to be imported at least from Spain and possibly from Arabia, as the foundation stallion for a breeding herd is also indicative of the financial status he expected of his readers.

18. Middle- and upper-class adolescents were also likely to spend their teen years away from their families, though more commonly in formal apprenticeships or at universities, in the case of males, or in the home of a family friend or female relative where they could learn the art of housewifery, in the case of females. The difference between education, whether academic, domestic, or professional, and service lies in the fact that young people sent out to be educated were supported by their own families, while servants had to earn their own wages and work for room and board.

19. Ann Kussmaul, *Servants in Husbandry in Early Modern England* (Cambridge, U.K.: Cambridge University Press, 1981), 31–35.

20. Ibid., 9.

21. Gervase Markham, *A Farewell to Husbandry; or, The inriching of all sorts of barren and sterill grounds in our kingedome, to be as fruitfull in all manner of pulse, and grasse as the best grounds whatsoever: together with the annoyances, and preservation of all graine and seede, from one yeare to many yeares. As also a husbandly computation of men and cattels dayly labours, their expences, charges, and uttermost profits. Attained by travel and experience, being a worke never before handled by any author and published for the good of the whole kingdome* (London: Printed by John Beale and Augustine Mathews, 1620, by M. Flesher for Richard Jackson, 1625, Nicholas Okes for John Harrison, 1631, Edward Griffin for John Harrison, 1638, William Wilson for John Harrison, 1649, and by E. O. for George Sawbridge, 1668).

22. Markham, *How to chuse*, 13^r.

23. Kussmaul, 35.

24. The other possibility is that they were following the ingredients of the master's feeding plan but were increasing the amounts. There is a good chance that this happened at least some of the time, because it would have brought credit to the servants without costing them anything.

25. Markham, *How to chuse*.

26. Markham would likely have been acquainted with the upper-class habit of directing servants to fulfill agricultural projects of all kinds, particularly equine ones from his social position as the son of a wealthy gentleman raised on a working estate and his exposure to other grand barns, including the royal stud farm and the royal carriage stables.

27. Fox has devoted a book and an edited collection to resolving the puzzle of oral culture and its relationship to print in early modern England. He has not, however, addressed the specific question of master-servant textual transmission. Fox's interpretation of the intersections between oral and print culture is in his *Oral and Literate Culture*. See also Adam Fox and Daniel Woulf, *The Spoken Word: Oral Culture in Britain, 1500–1850* (Manchester, U.K.: Manchester University Press, 2003).

28. The phrase appears in Latin within the border of the printer's colophon: "Tempore re patet occulta veritas." For a picture of this device and the story of its transfer to William Wood, see R. B. McKerrow, *Printers' and Publishers' Devices*, 181, device 312.

29. Markham, *How to chuse,* title page.
30. Ibid.
31. Ibid.
32. Kussmaul asserts that most servants moved on after only one year with any given family, both to broaden their base in occupational training and to expand their experiences in the world. After all, their freedom would be much more limited after marrying, and they were in service to acquire as much general knowledge as possible about the mechanics of good husbandry. Young men who were particularly adept with horses and proved successful at caring for them would probably have sought other jobs doing the same thing because those jobs tended to be better paid and carry more status. They also represented an entrée into the burgeoning equine industry. For the ways in which these two trends intersected, it is worth considering both Thirsk, 390–400, and Kussmaul, 31–35.
33. Jaggard 1599 edition, Fiv^r.
34. Markham, *How to chuse,* A2^v.
35. *Mirror of Alchimy,* 55.
36. Markham, *How to chuse,* C1^r.
37. "The admirable Force of Art and Nature," in *Mirror of Alchimy,* K2^v.
38. Markham, *How to chuse,* D1^r.
39. Jaggard 1599 edition, Aiii^r. For more on the use of authority in this text, see chapter 3.
40. Markham, *How to chuse,* B2^r.
41. Ibid., M1^r.
42. "The Mirror of Alchimy," in *Mirror of Alchimy,* B3^r.
43. Markham, *How to chuse,* B1^r.
44. "The Mirror of Alchimy" B1^v.
45. Ibid., B3^r.
46. Markham, *How to chuse,* ch. 1.
47. Ibid., B3^v.
48. Ibid., B3^r.
49. Ibid.
50. Ibid.
51. More will be said later about the cultural position of horse training in the sixteenth century and the social and economic classes who valued and participated in the equine industry, composed of horse owners, caretakers, riders, and trainers.
52. Markham, *How to chuse,* D2^r.
53. Ibid., M1^r.
54. The English inherited their model for good horsemanship and nice horses from the Spanish and Italians, and Markham's text is based on a combination of an Italian one called *The Art of Riding* and the instruction that his father, and possibly he, received at the hands of an Italian riding master hired by the court of Henry VIII, a tradition carried on by Elizabeth I (Thirsk, 387–92).
55. Markham, *How to chuse,* A4^v.
56. Ibid., B2^v.

57. Ibid., F4^v.

58. Ibid., M1^r, F4^r.

59. Nothing in early modern medicine was that easy to administer, particularly when the patient was likely to weigh up to a thousand pounds and dosing him required the removal of flaps of skin or teeth or the administration of a drench without the assistance of anesthesia.

60. Markham, *How to chuse,* M2^r.

61. Ibid., G1^r.

62. Ibid., M2^r.

63. Ibid., M1^r.

64. Ibid.

65. "First let him blood both in neck and temples, for the originall cause of a Feaver, is surfeit breeding putrifaction in the blood: then when hys shaking beginneth, take three newe layde Egges, breake them in a dishe and beate them together, them mixe thereto five or sixe spponefuls of excellent good Aqua vitae and give it him in a horne, then bridle him, and in some Close or Court, chase him till hi shaking cease and he beginne to sweath: then sette him up and clothe him warme" (Ibid., M2^v).

66. Ibid.

67. Ibid., A3^v.

68. Ibid.

69. Anonymous, "Airs, Waters, Places," in *The Hippocratic Writings,* edited by G. E. R. Lloyd (New York: Pelican Classics, 1978).

70. Markham, *How to chuse,* A3^r.

71. Ibid., A2^r.

72. Kussmaul, 35.

Conclusion

1. Roberts, "Shakespeare 'Creepes into the Women's Closets,'" 30–63.

2. Fox, *Oral and Literate Culture*; Watt, *Cheap Print.*

BIBLIOGRAPHY

Primary Sources

The boke of secretes of Albertus Magnus, of the vertues of Herbes, stones, and certayne beastes. Also, a boke of the same author, of the maruaylous thinges of the world: and of certaine effects caused of certaine beastes. London: John King, 1560. Tamer 862, Bodleian Library, Oxford University, Oxford, U.K. [Cited as King 1560 *Boke of secretes of Albertus Magnus.*]

The Boke of secretes of Albertus Magnus, of the vertues of herbes, stones, and certayne Beastes. Also, a boke of the same author, of the marvaylous thinges of the worlde: and of certain effectes, caused of certayne beastes. London: William Seres, 1570. Guildhall Library, London. [Cited as Seres 1570 *Boke of secretes of Albertus Magnus.*]

The booke of secretes of Albertus Magnus of the vertues of Herbes, stones and certayne Beastes. Also a booke of the same Author of the maruaylous thinges of the world, and of certayn effectes caused of certayne Beastes. London: William Copeland, 1565. Antiq.f.E91, Bodleian Library, Oxford University, Oxford, U.K. [Cited as Copeland 1565 *Booke of secretes.*]

Cornucopiae, Or diuers secrets: Wherein is contained the rare secrets in Man, Beasts, Foules, Fishes, Trees, Plantes, Stones and such like, most pleasant and profitable, and not before committed to bee printed in English. Newlie drawen out of diuers Latine Authors into English by Thomas Iohnson. London: Printed for William Barley, 1596. C143.b.20, Rare Books Collection, Bodleian Library, Oxford University, Oxford, U.K. [Cited as *Cornucopiae.*]

Markham, Gervase. *How to chuse, ride, traine, and diet, both Hunting-horses and running Horses. With all the secrets thereto belonging: an Arte never here-to-fore written by any Author. Also a discourse of horsemanship, wherein the breeding and ryding of Horses for service, in a briefe manner, is more methodically sette down then hath beene heretofore: with a more easie and direct course for the ignorant, to attaine to the said Arte or Knowledge. Together with a newe addition for the cure of horses diseases, of what kinde or natur soever.* London: By I. R. for Richard Smith, 1596. C.31.c.3, Humanities Collection, British Library, London. [Cited as Markham, *How to chuse.*]

The Mirror of Alchimy Composed by the thrice-famous and learned Fryer, Roger Bachon, sometimes fellow of Martin Colledge: and afterwards of Brasen-nose Colledge in Oxenforde. Also a most excellent and learned discourse of the admirable force and efficacie of Art and Nature written by the same Author. With certain other worthie treatises of the like Argument. London: Printed by Thomas Creede for Richard Olive, 1597. C.115.n.11, Humanities Collection, British Library, London. [Cited as *Mirror of Alchimy*.]

The Secrets of Albertus Magnus. Of the vertues of Hearbes, Stones, and certaine Beasts. Also a Booke of the same Author, of the maruelous things of the worlde, and of certaine effects caused by certaine beasts. London: William Jaggard, 1599. Shelf 146/A, Rare Books Collection, Wellcome Library, Wellcome Trust, London. [Cited as Jaggard 1599 *Secrets of Albertus Magnus*.]

The Treasurie of commodious Conceits, and hidden Secrets and may be called, The Huswives Closet, of healthfull provision. Mete and necessarie for the profitable use of all estates both men and women: And also pleasaunt for recreation. With a necessary Table of all things herein contained. Gathered at of sundrye Experiments lately practiced by men of great knowledge. London: Richard Jones, 1573. Call no. 59164, Rare Books Collection, Huntington Library, San Marino, Calif. [Cited as Partridge, *Treasurie of commodious Conceits*.]

The Treasurie of hidden Secrets. Commonlie called, The good Huswives Closet of Provision, for the health of her Houshold: Gathered out of sundry experiments, latelie practiced by men of great knowledge: And now newly inlarged, with divers necessarie Physicke helps, and knowledge of the names and naturall disposition of diseases, that most commonlie happen to men and women. Not impertinent for every good Huswife to use in her house, amongst her owne Famelie. London: Richard Jones, 1596. 1037.e2, Humanities Collection, British Library, London. A copy of the edition James Roberts printed for Edward White in 1600 is also at the British Library: call number 1037.e2. That copy is devoid of marginalia or any indication of reader use. [Cited as Jones 1596 *Treasurie of hidden Secrets*.]

The Widowes treasure, plentifully furnished with sundry precious and approved secretes in Phisicke, and Chirurgery for the health and pleasure of mankind. Hereunto are adioyned, sundrie pretie practices and conclusions of Cookerie: with many profitable and holesome medicines for sundrie diseases in Cattell. London: Printed by Edward Alde [*sic*] for Edward White, 1588. C.104.e.32(1–4), Humanities Section, British Library, London. The copy available at the British Library is actually the 1595 edition, a product of the partnership between James Roberts and Edward White discussed in chapter 1. The 1631 edition printed by Elizabeth Allde is available at the Bodleian Library, where it is packaged with the 1591 edition of *The Treasurie of commodious Conceits*. [Cited as *Widowes treasure*.]

Secondary Sources

Arber, Edward, ed. *A transcript of the registers of the Company of Stationers of London, 1554-1640.* 5 vols. London: privately printed, 1875; repr., Gloucester, Mass.: Peter Smith, 1967.
Armitage, David, ed. *Theories of Empire, 1450-1800.* Aldershot, U.K.: Ashgate, 1998.
Armstrong, Nancy, and Leonard Tennenhouse. *The Ideology of Conduct: Essays on Literature and the History of Sexuality.* New York: Methuen, 1987.
Assarson-Rizzi, Kerstin. *Friar Bacon and Friar Bungay: A Structural and Thematic Analysis of Robert Greene's Play.* Berlingska: Lund, 1972.
Bacon, Roger. *Opera quaedam hactenus inedita,* edited by J. S. Brewer. London: n.p, 1859.
Barthes, Roland. *S/Z,* trans. Richard Miller. New York: Hill & Wang, 1970.
Beier, A. L. *Masterless Men: The Vagrancy Problem in England, 1550-1640.* London: Methuen, 1985.
Brehm, Edmund. "Roger Bacon's Place in the History of Alchemy," *Ambix* 23,1 (1976): 53-58.
Brooks, Douglas. *From Playhouse to Printing House: Drama and Authority in Early Modern England.* New York: Routledge, 2000.
Browne, Pamela. "Laughing at the Cony: A Female Rogue and 'The Verdict of the Smock.'" *English Literary Renaissance* 29,2 (1999): 201-24.
Burke, Peter. *Popular Culture in Early Modern Europe.* New York: New York University Press, 1978.
Burns, William E. "'A Proverb of Versatile Mutability': Proteus and Natural Knowledge in Early Modern Britain." *Sixteenth Century Journal* 32,4 (2001): 969-80.
Capp, Bernard. "Separate Domains? Women and Authority in Early Modern England." In *The Experience of Authority in Early Modern England,* edited by Paul Griffiths, Adam Fox, and Steve Hindle, 117-45. New York: St. Martin's Press, 1996.
Chartier, Roger. *The Cultural Use of Print in Early Modern France,* translated by Lydia G. Cochrane. Princeton, N.J.: Princeton University Press, 1987.
———. "Popular Appropriations: Readers and Their Books." In *Forms and Meanings: Texts, Performances, and Audiences from Codex to Computer,* edited by Roger Chartier, 83-98. Philadelphia: University of Pennsylvania Press, 1995.
Clark, Stuart. *Thinking with Demons: The Idea of Witchcraft in Early Modern Europe.* Oxford, U.K.: Oxford University Press, 1997.
Copenhaver, Brian. "Natural Magic, Hermetism, and Occultism in Early Modern Science." In *Reappraisals of the Scientific Revolution,* edited by David C. Lindberg and Robert S. Westman, 261-301. Cambridge, U.K.: Cambridge University Press, 1990.
Cox, Richard. *Hibernia Anglicana or the History of Ireland from the Conquest Thereof by the English to this Present Time.* London: Printed by H. Clark for Joseph Watts, 1692.

Crawford, Patricia. *Women and Religion in England, 1500–1720*. London: Routledge, 1993.
Crawford, Patricia, and Laura Gowing. *Women's Worlds in Seventeenth Century England*. New York: Routledge, 2000.
Cressy, David. *Travesties and Transgressions in Early Modern England*. Oxford, U.K.: Oxford University Press, 2000.
Crombie, A. C., and J. D. North. "Bacon, Roger." In *Dictionary of Scientific Biography*, edited by Charles Gillispie. New York: Oxford University Press, 1970–90.
Daston, Lorraine. History of Science Society Distinguished Lecture at the annual meeting of the History of Science Society. Milwaukee, Wisc., Nov. 18–22, 2002.
———. "The Nature of Nature in Early Modern Europe." *Configurations* 6 (1998): 149–72.
Dean, Paul. "'Friar Bacon and Friar Bongay' and 'John of Bordeaux': A Dramatic Diptych." *English Language Notes*, 18,4 (1981): 262–66.
———. "Shakespeare's Henry VI Trilogy and Elizabethan Romance Histories: The Origins of a Genre." *Shakespeare Quarterly* 33,1 (1982): 34–48.
De Grazia, Margaret, Maureen Quilligan, and Peter Stallybrass, eds. *Subject and Object in Renaissance Culture*. Cambridge, U.K.: Cambridge University Press, 1996.
Duff, Edward Gordon. *A Century of the English Book Trade: Short Notices of All Printers, Book-Binders, and Others Connected with It from the Issue of the First Dated Book in 1457 to the Incorporation of the Company of Stationers in 1557*. London: Printed for the Bibliographical Society by Blades, East, and Blades, 1905.
Dutton, Richard. *Mastering the Revels: The Regulation and Censorship of English Renaissance Drama*. Iowa City: Iowa University Press, 1991.
Eamon, William. *Science and the Secrets of Nature: Books of Secrets in Medieval and Early Modern Culture*. Princeton, N.J.: Princeton University Press, 1994.
Easton, Stewart. *Roger Bacon and His Search for a Universal Science*. Oxford, U.K.: Oxford University Press, 1962.
Eisenstein, Elizabeth. *The Printing Press as an Agent of Change: Communication and Cultural Transformation in Early Modern Europe*. Cambridge, U.K.: Cambridge University Press, 1979.
Engelsing, Rolf. *Der Burger als Leser, Lesergeschichte in Deutschland, 1500–1800*. Stuttgart, Ger.: Metzlersche, 1974.
Fehrenbach, R. J., ed. *Private Libraries in Renaissance England: A Collection and Catalogue of Tudor and Early Stuart Book-lists*. Binghamton: Medieval and Early Modern Renaissance Texts and Studies, 1992.
Ferguson, Margaret. *Dido's Daughters Literacy, Gender, and Empire in Early Modern England and France*. Chicago: University of Chicago Press, 2003.
Findlen, Paula. "Empty Signs? The Book of Nature in Renaissance Science." *Studies in History and Philosophy of Science* 21,3 (1990): 511–18.
———. "Science as a Career in Enlightenment Italy: The Strategies of Laura Bassi." *Isis* 84,3 (1993): 440–69.
Fissell, Mary. "Hairy Women and Naked Truths: Gender and the Politics of Knowledge in *Aristotle's Masterpiece*." *William and Mary Quarterly* 60,1 (2003): 43–74.

———. "Readers, Texts, and Contexts: Vernacular Medical Works in Early Modern England." In *The Popularization of Medicine, 1650–1850*, edited by Roy Porter, 72–96. London: Routledge, 1992.

Fox, Adam. *Oral and Literate Culture in England, 1500–1700*. Oxford, U.K.: Oxford University Press, 2000.

Fumerton, Patricia. "Not Home: Alehouses, Ballads, and the Vagrant Husband in Early Modern England." *Journal of Medieval and Early Modern Cultural Studies* 32,3 (2002): 498–518.

Goldberg, Jonathan. *Endlesse Worke: Spenser and the Structures of Discourse*. Baltimore: Johns Hopkins University Press, 1981.

Gowing, Laura. *Domestic Dangers: Women, Words, and Sex in Early Modern London*. Oxford, U.K.: Oxford University Press, 1999.

Greenblatt, Steven. "Toward the Poetics of Culture." In *The New Historicism*, edited by H. Aram Veeser, 1–14. New York: Routledge, 1989.

Greene, Robert. *The blacke bookes messenger: Laying open the life and death of Ned Browne one of the most notable cutpurses, crosbiters, and conny-catchers, that ever lived in England*. London: Printed by John Danter for Thomas Nelson, 1592.

———. *The defence of conny-catching*. London: Printed by A. Jeffes for Thomas Gubbins, 1592.

———. *A disputation between a hee conny-catcher and a shee conny-catcher: whether a theefe or a whoore is most hurtfull in cousonage, to the common-wealth. Discovering the secret villainies of alluring strumpets*. London: A. Jeffes for T. Gubbin, 1592.

———. (1594). *The honorable historie of frier Bacon and frier Bongay*. London: Privately printed for Edward White, 1594.

———. *A notable discovery of coosenage, now daily practiced by Sundry lewd persons called connie-catchers and crosse-byters*. London: Printed by John Wolfe for Thomas Newman, 1591.

———. *The Second part of conny-catching*. London: Printed by John Wolfe for William Wright, 1591.

———. *The third and last part of conny-catching*. London: Printed by Thomas Scarlet for Cuthbert Burby, 1592.

Greg, Walter M., ed. *Henslowe's Diary*. Part 1, *Text*. London: A. H. Bollen, 1904.

Grosart, Alexander B., ed. *The Life and Works in Prose and Verse of Robert Greene, MA*. 15 vols. London: for private circulation only, 1881–86.

Hall, David. *Cultures of Print: Essays in the History of the Book*. Amherst: University of Massachusetts Press, 1996.

Hatfield, Andrew. *Literature, Politics, and National Identity*. Cambridge, U.K.: Cambridge University Press, 1994.

Hausknecht, Gina. "'So Many Shipwracke for Want of Better Knowledge': The Imaginary Husband in Stuart Marriage Advice." *Huntington Library Quarterly* 64,1–2 (200):81–106.

Heilbron, J. L. *Electricity in the Seventeenth and Eighteenth Centuries: A Study of Early Modern Physics*. Berkeley: University of California Press, 1979.

Helgerson, Richard. *Forms of Nationhood: The Elizabethan Writing of England*. Chicago: University of Chicago Press, 1992.
Hieatt, Charles. "A New Source for Friar Bacon and Friar Bongay." *Review of English Studies* 32, 26 (1981): 180-87.
Kavey, Allison. "Mercury Falling: Gender Flexibility and Eroticism in Popular Alchemy." In *The Sciences of Homosexuality in Early Modern Europe*, edited by George Rousseau and Kenneth Borris. New York: Routledge, 1998.
Knowles, James. (1998). "'Infinite Riches in a Little Room': Marlowe and the Aesthetics of the Closet." In *Renaissance Configurations—Voices/Bodies/Spaces, 1580-1690*, edited by Gordon McMullan, 1-17. London: Macmillan, 1998.
Kuhn, Thomas. *The Structure of Scientific Revolutions*. Chicago: University of Chicago Press, 1996.
Kussmaul, Ann. *Servants and Husbandry in Early Modern England*. Cambridge, U.K.: Cambridge University Press, 1981.
Levin, J. A., ed. *Friar Bacon and Friar Bungay*. London: Ernest Benn, 1969.
Levin, Richard. "Tarlton in *The Famous History of Friar Bacon and Friar Bongay*." *Medieval and Renaissance Drama* 12 (1999): 84-98.
Lindberg, David. "Lines of Influence in Thirteenth-Century Optics: Bacon, Witelo, and Pecham." *Speculum* 46 (1971): 72-75.
———. *Roger Bacon's Philosophy of Nature: A Critical Edition, with English Translation, Introduction, and Notes, of De multiplicatione specierum and De speculis comburentibus*. Oxford, U.K.: Clarendon Press, 1983.
Linden, Stanton J., ed. *The Mirror of Alchimy: Composed by the Thrice-Famous and learned fryer, Roger Bacon*. New York: Garland Press, 1992.
Lloyd, G. E. R., ed. *The Hippocratic Writings*. New York: Pelican, 1978.
Loewenstein, Joseph. "For a History of Literary Property: John Wolfe's Reformation." *English Literary Renaissance* 18,3 (1988): 389-412.
———. "The Script and the Marketplace." *Representations* 12 (1985): 101-14.
Long, Pamela O. *Openness, Secrecy, Authorship: Technical Arts and the Culture of Knowledge from Antiquity to the Renaissance*. Baltimore: Johns Hopkins University Press, 2001.
Mack, Phyllis. *Visionary Women: Ecstatic Prophecy in Seventeenth-Century England*. Berkeley: University of California Press, 1992.
Markham, Gervase. *Cavelarice; or, The English Horseman*. London: Printed by Edward Allde and William Jaggard for Edward White, 1607.
McKerrow, R. B. *A Dictionary of Printers and Booksellers in England, Scotland, and Ireland, and of Foreign Printers of English Books 1557-1640*. London: Printed for the Bibliographic Society by Blades, East, and Blades, 1910.
———. *Printers' and Publishers' Devices in England and Scotland, 1485-1640*. London: Printed for the Bibliographical Society at the Chiswick Press, 1913.
McMillin, Scott. (1984). "The Queen's Men in 1594: A Study of 'Good' and 'Bad' Quartos." *English Literary Renaissance* 14,1 (1984): 55-69.
McMullan, Gordon. *Renaissance Configurations—Voices/Bodies/Spaces, 1580-1690*. London: Macmillan, 1998.

Mendelson, Sara. *The Mental World of Stuart Women: Three Studies.* Brighton, U.K.: Harvester Press, 1987.
Mendelson, Sara, and Patricia Crawford. *Women in Early Modern England, 1550–1720.* Oxford, U.K.: Clarendon Press, 1998.
Miller, D. A. *The Novel and the Police.* Berkeley: University of California Press, 1998.
Montrose, Louis. "Spenser's Domestic Domain: Poetry, Property, and the Early Modern Subject." In *Subject and Object in Renaissance Culture,* edited by Margreta De Grazia, Maureen Quilligan, and Peter Stallybrass, 83–132. Cambridge, U.K.: Cambridge University Press, 1996.
Mueller, Judith C. "Fallen Man: Representations of Male Impotence in Britain." *British Studies in Eighteenth Century Culture* 28 (1998): 85–102.
Murphy, Andrew. *But the Irish Sea betwixt Us.* Lexington: University of Kentucky Press, 1999.
Nicholl, Charles. *The Alchemical Theater* London: Routledge, 1980.
Nummedal, Tara E. "Alchemical Reproduction and the Career of Anna Maria Zieglerin." *Ambix* 48,2 (2001): 56–68.
Orgel, Steven. *The Authentic Shakespeare and Other Problems of the Early Modern Stage.* New York: Routledge, 2002.
Palmer, Patricia Ann. *Language and Conquest in Early Modern Ireland: English Imperial Literature.* Cambridge, U.K.: Cambridge University Press, 2001.
Partridge, John. *A Treasurie of Commodious Conceits and Hidden Secrets.* London: Printed for Richard Jones, 1573.
———. *A Treasurie of Hidden Secrets.* London: Printed for Richard Jones, 1591.
Pelling, Margaret. "The Women of the Family? Speculations around Early Modern British Physicians." *Social History of Medicine* 8,3 (1995): 383–401.
Pennell, Sara. "Pots and Pans History: The Material Culture of the Kitchen in Early Modern England." *Journal of Design History* 11,3 (1998): 201–16.
Petersen, Douglas L. "Lyly, Greene, and Shakespeare and the Recreations of Princes." *Shakespeare Studies* 20 (1987): 67–88.
Porter, Roy. "'The Secrets of Generation Display'd': *Aristotle's Masterpiece* in Eighteenth-Century England." *Eighteenth Century Life* 9,3 (1985): 1–21.
Pratchett, Terry. *Hogfather.* New York: Harper Collins, 1996.
Principe, Lawrence M. "Robert Boyle's Alchemical Secrecy: Codes, Ciphers, and Concealments." *Ambix* 39,2 (1992): 63–74.
Pritchard, Alan. "Thomas Charnock's Book Dedicated to Queen Elizabeth." *Ambix* 26,1 (1979): 56–72.
Pseudo-Bacon, Roger. *Frier Bacon His Discovery of the Miracles of Art, Nature, and Magick.* London: Printed by Simon Miller, 1659.
———. *The Mirror of Alchimy.* London: Printed for Richard Olive, 1597.
Pugliatti, Paola. *Beggary and Theatre in Early Modern England.* Aldershot, U.K.: Ashgate, 2003.
Roberts, Julian, and Andrew G. Watson, eds. *John Dee's Library Catalogue.* London: London Bibliographic Society, 1990.

Roberts, Sasha. "Shakespeare 'Creepes into the Women's Closets about Bedtime': Women Reading in a Room of their Own." In *Renaissance Configurations—Voices/Bodies/Spaces, 1580–1690,* edited by Gordon McMullan, 30–63. London: Macmillan, 1998.

Sadler, Lynn Veach. "Alchemy and Greene's *Friar Bacon and Friar Bungay.*" *Ambix* 22,2 (1975): 111–24.

Sedgwick, Eve Kosofsky. *Epistemology of the Closet.* Berkeley: University of California Press, 1991.

Shapin, Steven. *A Social History of Truth: Civility and Science in Seventeenth-Century England.* Chicago: University of Chicago Press, 1994.

Spufford, Margaret. *Small Books and Pleasant Histories: Popular Fiction and Its Readership in Seventeenth Century England.* Athens: University of Georgia Press, 1982.

Stallybrass, Peter. "Worn Worlds." In *Subject and Object in Renaissance Culture,* edited by Margreta De Grazia, Maureen Quilligan, and Peter Stallybrass, 1–16. Cambridge, U.K.: Cambridge University Press, 1996.

Stavreva, Kirilka. "Scaffolds unto Prints: Executing the Insubordinate Wife in the Ballad Trade of Early Modern England." *Journal of Popular Culture* 31,1 (1997): 177–88.

Stine, Jennifer L. *Opening Closets: The Discovery of Household Medicine in Early Modern England.* Ph.D. dissertation, Oxford University, 1996. Dissertation Abstracts International 57(5): 2169–A.

Tetzeli von Rosador, Kurt. "The Sacralizing Sign: Religion and Magic in Bale, Greene, and Early Shakespeare." *Yearbook of English Studies* 23 (1993): 30–45.

Thirsk, Joan, ed. *The Rural Economy of England.* London: Hambledon Press, 1984.

Walsham, Alexandra. *Providence in Early Modern England.* Oxford, U.K.: Oxford University Press, 1999.

Watt, Tessa. *Cheap Print and Popular Piety.* Cambridge, U.K.: Cambridge University Press, 1991.

Webb, Peter Saunders. *Writing in the Bearpit: Popular Authors in Early Modern England.* Ph.D. dissertation, University of Michigan, 1991.

Weil, Rachel. *Political Passions: Gender, the Family, and Political Argument in England, 1680–1714.* New York: St. Martin's Press, 1999.

Wiltenberg, Joy. *Disorderly Women and Female Power in the Street Literature of Early Modern England and Germany.* Charlottesville: University of Virginia Press, 1992.

Yates, Frances. (1979). *The Occult Philosophy in the Elizabethan Age.* London: Routledge, 1979.

Zamelli, Paola. *The Speculum Astronomiae and Its Enigma: Astrology, Theology, and Science in Albertus Magnus and His Contemporaries.* Boston: Kluver Academic, 1992.

INDEX

Admirable Force of Art and Nature, The, 52
Albertus Magnus, 4–5, 11, 21, 57
Alchemy, 4, 35, 44 85, 153; and gendered knowledge, 98–99, 119; and pseudo-Roger Bacon, 53–56
Allde, Edward, 17, 19–22, 52, 58–59, 95
antipathy, 6, 61–62, 84–88, 91, 93, 156
Aristotle's Masterpiece, 32–34
Astrology, 1, 37, 70–71, 101; in *The Secrets of Albertus Magnus*, 75–78, 85; in *Treasurie of Hidden Secrets*, 121–22
audience. *See* readers
authority, 1, 3–5, 32–35, 48–51, 56–57, 72, 74, 101, 103–5, 120–22, 140–41, 148–51

Bacon, Roger, 36–37
Barley, William, 71–73
Bongay, Friar, 42, 80, 87
Book of Knowledge, 1, 32, 35

Charlewood, John, 20–21, 23, 26
cheap print, 2, 15–16, 18, 31, 157
Copeland, William, 10–13, 16, 24, 67–69
Cornucopiae or divers secrets, 4; and comparisons with *Albertus Magnus*, 10, 59–94; and printer strategies, 71–73; and sympathy and antipathy, 87–88; and textual structure, 65–66, 82–84, 89–92

Dee, John, 16, 35, 38, 42, 57
distillation, 54–57, 98, 119–20

Emerald Table, 48–49, 53
Erra Pater, 1, 32–33, 35

femininity, 7–8, 75–77, 96–97, 98–104, 113, 116–17, 124–25
Franciscans, 73–75

Greene, Robert, 6, 22–29, 38–40, 57

honorable historie of friar Bacon and friar Bongay, The, 6, 38–47; and authority, 44–45; and nationalism, 45–47; and relationship between art and magic, 42–47; and secrets and gender, 42–44; and systems of sympathy in, 41
horses, in early modern England, 127–31; and gentlemen, 131–32; and training, 144–47
Hortulanus, 48–49, 53
How to chuse, ride, traine, and diet both Hunting-horses, With all the secrets thereto belonging discovered: an Arte never here-to-fore written by

any Author, 126–55; audience characteristics, 133–35, 137–39; authority in, 140–42, 150–49, 152; comparison to *Mirror of Alchimy*, 139–43; relationship to secrecy, 135–37; recipes, 144–45; veterinary knowledge within, 148–50

Jaggard, William, 14–17, 60, 67, 69–72, 78
Jones, Richard, 17–21, 95–96, 102, 106–8, 124

King, John, 10–11, 16

lodestone, 75–76, 79, 89–91

magic, 40, 42–43, 45–47, 50, 58, 74
Markham, Gervase, 4, 8, 21; and constructions of masculinity, 127, 134–35, 142–47, 153; and constructions of readers, 132–35, 137–39; relationship to other books of secrets, 139–45
masculinity, 8, 12, 127, 134–35, 142–47, 153
medicine, 4, 7, 11, 14, 74, 98, 100, 105–6; and human, 34, 74, 105–7, 110–17, 120–21; and humoral, 33–34, 115, 118–19; and recipes, 110–15, 118, 144–46; and veterinary, 112, 117, 148–50
Mirror of Alchimy, The, 3–4, 7; and comparisons *How to chuse, ride, diet, and train, both Hunting-horses and running Horses*, 139–43; and constructions of authority, 48–51; and good readers of, 51–54; and models for natural world, 54–55; and obscurity and secrecy in, 51–53; and practical knowledge, 55–56; and pseudo-Roger Bacon, 47–50, 91, 93

Partridge, John, 4, 95–96, 103–4, 108
Pliny, 34, 51, 57, 93

print marketplace, 9–31
printer strategies, and within *Albertus Magnus*, 66–71, 92; and within *Cornucopiae or divers secrets*, 71–72, 79–82, 92; and within *Treasurie of commodious conceits and hidden secrets*, 100–104; and within the *Treasurie of hidden secrets*, 106–8; within *The Widowes Treasure*, 104–6
pseudo-Bacon, Roger, 3–4, 27–28, 35, 47–58

readers, 2, 5, 7–8, 31, 52–56, 62, 72–78, 83–84, 93, 107–8, 156–57; and female, 7, 20, 98–99, 114–20, 125; and male, 8, 127, 131–35, 137–38, 141–42
reading, intensive, 52–53, 64–65
Roberts, James, 20–21, 187

Secrets of Albertus Magnus, The, 3, 7,9, 29, 51; and astrology, 77–78; and comparisons with *Cornucopiae, or divers secrets*, 10, 59–94; and printers and printer relationships, 10–17, 61; and readers and experience, 72–79, 94; and sympathy and antipathy, 84–87; and textual structure, 60–72, 79–82, 88–92
Seres, William, 13–14, 59–61
Stationers' Company, 6, 9–31
sympathy, 61–62, 84–86, 93, 156

Treasurie of commodious conceits and hidden secrets, The, 7, 9, 19, 95–96; and comparison with *Treasurie of Hidden Secrets*, 104–12; and comparison with *Widowes Treasure*, 104–13; and gendered spaces and knowledge, 113–15; and medicine, human, 120–22; and structure, 98–111
Treasurie of Hidden Secrets, The, 6, 19–21, 96; and comparison with *Widowes Treasure*, 104–13; and gen-

dered spaces and knowledge, 117–20;
and medicine, human, 120–22; and
structure, 98–111
Trismegistus, Hermes, 35, 48–50, 53

White, Edward, 17, 20–28
Widowes Treasure, The, 6, 9, 14,
17, 95–96; and comparison with
*Treasurie of Commodious Conceits
and Hidden Secrets and Treasurie of
Hidden Secrets,* 104–13; and gendered spaces and knowledge, 116–17;
and medicine, human, 116–17; and
medicine, veterinary, 117; and reader
characteristics, 105–6; and structure,
109–10
witnessing, 34, 83–84, 148–51
women, 7; and natural knowledge,
96–100; and sexuality, 75–77

ALLISON KAVEY is an assistant professor of history at CUNY John Jay College of Criminal Justice. She has contributed to *The Sciences of Homosexuality in Early Modern Europe*, edited by George Rousseau and Kenneth Borris.

The University of Illinois Press
is a founding member of the
Association of American University Presses.

Composed in 11/14 Bulmer
with Woodtype Ornaments display
by Jim Proefrock
at the University of Illinois Press
Designed by Paula Newcomb
Manufactured by Thomson-Shore, Inc.

University of Illinois Press
1325 South Oak Street
Champaign, IL 61820-6903
www.press.uillinois.edu